MINI

ENCYCLOPEDIA

EARTH

ENCYCLOPEDIA

EARTH

Authors
John Farndon
Steve Parker

Miles
KeLLy

First published in 2015 by Miles Kelly Publishing Ltd
Harding's Barn, Bardfield End Green, Thaxted, Essex, CM6 3PX

Copyright © Miles Kelly Publishing Ltd 2015

This edition printed 11/16

LOT#:
2 4 6 8 10 9 7 5 3 1

Publishing Director Belinda Gallagher
Creative Director Jo Cowan
Editorial Director Rosie Neave
Designer Simon Lee
Image Manager Liberty Newton
Indexer Marie Lorimer
Production Elizabeth Collins, Caroline Kelly
Reprographics Stephan Davis, Jennifer Cozens, Thom Allaway
Assets Lorraine King

ISBN: 978-1-78617-174-0

Printed in China

British Library Cataloging-in-Publication Data
A catalog record for this book is available from the British Library

Made with paper from a sustainable forest

www.mileskelly.net

Contents

Earth in space and time

Formation of the Earth	12	The Ages of the Earth	22
Earth in space	14	Ice ages	24
Earth and Moon	16	Between ice ages	26
Shape of the Earth	18	Fossils	28
Scanning the Earth	20		

Structure of the Earth

The Earth's chemistry	32	Lithosphere	52
Minerals	34	Continental drift	54
Rocks	36	Tectonic plates	56
Igneous rocks	38	Converging plates	58
Metamorphic rocks	40	Diverging plates	60
Sedimentary rocks	42	Faults	62
Earth's interior	44	Folds	64
Earth's core	46	High mountains	66
The mantle	48	Mountain ranges	68
The crust	50		

Volcanoes and earthquakes

Volcanoes	72	Hot-spot volcanoes	80
Types of volcano	74	Earthquakes	82
Volcanic eruptions	76	Earthquake damage	84
Lava and ash	78	Earthquake measurement	86
		Earthquake waves	88
		Earthquake prediction	90
		Famous earthquakes	92
		Tsunamis	94

Shaping the land

Changing landscapes	98	Swamps and marshes	118
Weathering	100	Lakes	120
Limestone weathering	102	Deserts	122
Caves	104	Glaciers	124
Rivers	106	Glaciated landscapes	126
Hills	108	Cold landscapes	128
River valleys	110		
River channels	112		
Waterfalls	114		
Floods	116		

Continents and oceans

Earth's continents	132	Southern Ocean	162
Europe	134	Seas	164
Africa	136	Tides	166
Asia	138	Waves	168
Australia, island continent	140	Beaches	170
Oceania	142	Rocky coasts	172
Antarctica	144	Coral reefs	174
North America	146	Icebergs	176
South America	148	Ocean deeps	178
Major islands	150	Black smokers	180
"World Ocean"	152	Surface ocean currents	182
Atlantic Ocean	154	Deep ocean currents	184
Arctic Ocean	156	Ocean trenches	186
Indian Ocean	158	Mariana Trench	188
Pacific Ocean	160	Puerto Rico Trench	190

Mountains and canyons

Mountain systems	194	Atlas	208
Himalayas	196	Transantarctic Mountains	210
Andes	198	Mid-Atlantic Ridge	212
The Rockies	200	Grand Canyon	214
Urals	202	Indus River Gorge	216
European Alps	204	Other canyons and gorges	218
Great Dividing Range	206		

Great rivers and lakes

Rivers through the ages	222	How lakes form	238
Amazon	224	Great Lakes	240
Nile	226	Lake Baikal	242
Mississippi	228	Lake Victoria	244
Congo	230	Lake Tanganyika	246
Yangtze	232	Great Bear Lake	248
Ganges	234	Lake Vostok	250
Yenisei	236		

Atmosphere and weather

Atmosphere	254	Snow and hail	268
Lights in the sky	256	Ice and cold	270
Sunshine	258	Air pressure	272
Air moisture	260	Weather fronts	274
Clouds	262	Wind	276
Fog and mist	264	Thunderstorms	278
Rain	266	Dust and sand storms	280
		Blizzards and whiteouts	282
		Tornadoes	284
		Hurricanes	286
		Seasons	288
		Climate zones	290
		Weather forecasting	292

Exploiting Earth

Earth's riches	296	Renewable energy	306
Surveying and prospecting	298	Freshwater	308
Metals and minerals	300	Farming the land	310
Gems and jewels	302	Exploiting the seas	312
Fossil fuels	304		

Living Earth

Ecosystems	316	Grasslands	330
Biomes	318	Life in dry lands	332
Biodiversity	320	Wetland and freshwater	334
Tropical forests	322	Freshwater life	336
Temperate forests	324	Coastal habitats	338
Boreal forests	326	Life in the open ocean	340
Polar and mountain life	328	Deep sea life	342

Earth in danger

Global warming	346	Threatened species	360
Climate change	348	Conservation	362
Pollution	350	Future Earth	364
Acid and ozone	352		
Drought and desertification	354		
Overfarming	356	Index	368
Vanishing resources	358	Acknowledgments	383

Earth in space and time

Formation of the Earth

- **The Earth formed 4.57 billion years ago**, probably out of debris left over from the explosion of a giant star.

- **Star debris** spun round the newly formed Sun and clumped into rocks called planetesimals.

- **Planetesimals** were pulled together by their own gravity to form planets including Earth and Mars.

- **At first**, the Earth was a seething mass of molten rock.

- **After 50 million years** a giant rock collided with the newborn Earth. The impact threw out debris, which gradually joined together to become the Moon.

- **The shock of the impact** that formed the Moon made iron and nickel collapse toward the Earth's center. They formed a core so dense that its atoms fused in nuclear reactions that have kept the inside of the Earth hot.

- **Molten rock** formed a mantle about 1,800 mi thick around the metal core. The core's heat keeps the mantle warm and churning, like boiling treacle.

1

- **The surface cooled and hardened** to form a thin crust. Then the Late Heavy Bombardment (LHB) of meteorites and asteroids about 3.9 billion years ago smashed the crust.

- **Steam and gases** billowing from volcanoes formed the Earth's first, poisonous atmosphere. The steam condensed to water.

- **By 3.8 billion years ago** the crust cooled again, and land and oceans formed. One Earth day was about 15 hours long.

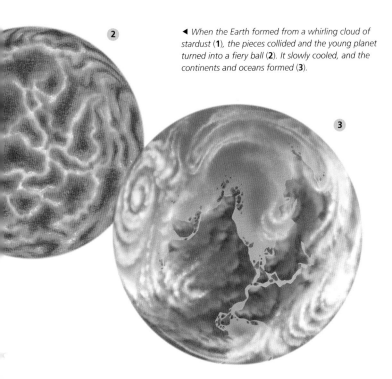

◄ When the Earth formed from a whirling cloud of stardust (**1**), the pieces collided and the young planet turned into a fiery ball (**2**). It slowly cooled, and the continents and oceans formed (**3**).

Earth in space

- **The Earth moves around**, or orbits, the Sun at an average speed of 18.5 mi/s.

- **The Earth completes** one orbit around the Sun approximately every 365 days or one year. It spins on its axis about once every 24 hours or one day.

- **Of the eight planets** in the Solar System, the Earth is the fourth largest, and it is the third-closest to the Sun.

- **Four of the eight planets** in the Solar System are terrestrial, or rocky, planets: Mercury, Venus, Earth, and Mars. Earth is the largest of these.

- **The Earth is the most dense** planet in the Solar System, at an average of 3.2 oz per cu in.

- **There are bodies**, known as quasi-satellites, including 2002 AA29 and Cruithne, that appear to orbit the Earth, in fact they follow the Earth around the Sun on a similar orbit.

- **On average**, the Earth is 93 million mi from the Sun—just the right distance to have suitable temperatures and liquid water for life to exist. It is nicknamed the "Goldilocks Planet" and is the only planet so far discovered with life.

- **Earth has seasons** because its axis is tilted by 23.4 degrees, meaning different areas of it are nearer the Sun, and so warmer, at different times of the year.

▶ *As spacecraft orbit the Moon, they see the Earth rise above the horizon. As the Earth spins different parts of it are lit by the Sun, and experience daytime.*

DID YOU KNOW?

From space, the Earth appears as "the blue planet" as most of it is covered by liquid water.

Earth and Moon

- **The Earth has only one natural satellite**, the Moon, which is the largest satellite, compared to its planet, in the Solar System.

- **The Moon is about one-quarter** of the diameter of the Earth and $\frac{1}{80}^{th}$ of the mass. Some astronomers regard the two as a double planet system.

▼ *The Moon's orbit (**1**) around the Earth is elliptical or oval, being around 225,500 mi at its nearest point and 252,000 mi at its farthest. Likewise Earth's orbit around the Sun (**2**) varies from 91–94 million mi.*

Sun

Earth

Moon

- **The Moon orbits** the Earth at an average distance of 238,855 mi and it is getting farther away by about 1.5 in every year.

- **Gravity on the Moon** is much weaker than on Earth so everything has only one-sixth the weight, or more correctly the mass, than on Earth.

- **However, the Moon's gravity** is slowing the Earth's rotation by approximately 2.3 milliseconds per day per century, so that a leap second has to be added to a year every so often.

- **The gravity of the Moon** is also enough to pull the oceans' water into bulges that move round the Earth as it rotates, which creates the tides.

- **The Moon rotates** at the same speed as it orbits the Earth so the same side always faces us, the far side has only ever been seen from spacecraft.

- **Although the Moon appears** to shine with a silvery light, this is only reflected sunlight—it does not produce light of its own.

- **The Moon orbits** the Earth every 27.3 days, and the part that reflects the Sun's light changes from new Moon to full Moon. This is called the phases of the Moon.

- **From Earth**, the Moon looks almost the same size as the Sun, which allows it to cover all of the Sun at certain times. This is called a total solar eclipse.

Shape of the Earth

▲ *The ancient Greeks realized that the Earth is a globe. Satellite measurements show that it is not quite perfectly round.*

● **The study of the shape** of the Earth is called geodesy. In the past, geodesy depended on ground-based surveys. Today, satellites play a major role.

● **The Earth is not a perfect sphere**. It is a unique shape called a geoid, which means "Earth-shaped."

- **At the Equator**, the Earth spins faster than at the Poles. This is because the Equator is farther from the Earth's spinning axis.

- **The extra speed** at the Equator pushes the Earth out in a bulge, while it is flattened at the Poles.

DID YOU KNOW?

A person standing at the Equator is actually moving at a speed of 1,038 mph as the Earth spins on its axis.

- **This Equatorial bulge** was predicted in 1687 by English scientist Isaac Newton (1642–1727).

- **The Equatorial bulge** was confirmed 70 years after Newton, by French surveys in Peru by explorer and mathematician Charles de La Condamine (1701–1774), and in Lapland by mathematician and philosopher Pierre de Maupertuis (1698–1759).

- **The Earth's diameter** at the Equator is almost 8,000 mi. This is longer—by 27 mi—than the vertical diameter from the North Pole to the South Pole.

- **The official measurement** of the Earth's radius at the Equator is 3,963 mi.

- **The LAGEOS** (Laser Geodynamic) series of satellites, first launched in 1976, reflect laser beams to make very accurate measurements of the Earth. They can measure movements of the Earth's tectonic plates as small as an inch.

- **The Seasat satellite** of 1978 confirmed the ocean surfaces are geoid. It took millions of measurements of the height of the ocean surface, accurate to within a few inches.

Scanning the Earth

● **Satellites in space** reveal a lot about the Earth. The TOPEX/Poseidon satellite measured the height of the ocean surface to a few inches. This accuracy revealed mountains on the seabed because their extra gravity creates a slight difference in the height of the water.

▼ *A view of Iran from the Landsat 7 satellite. It was built up and colored by computer from multiple photos taken using infrared light to get a clear view through clouds.*

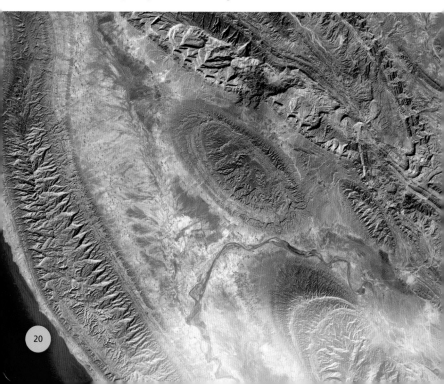

- **The pull of gravity** varies slightly over the Earth. The twin GRACE satellites can detect these variations accurately, revealing a lot about the Earth's oceans and interior.

- **NASA's Aqua satellite** can detect how moist soil is and so tell farmers if crops need water, or when the soil is ready for planting.

- **Satellite images** can be used to find mineral deposits and identify earthquake-prone areas. They can also chart the flow of the Gulf Stream and other currents that affect weather and climate.

- **Satellite pictures** help us to gauge how human activities are affecting the planet. They monitor ocean temperatures, measure the warming of the atmosphere and document deforestation.

- **Satellites reveal geology** and landforms where oil or minerals might be. Satellites have shown deposits of copper, nickel, zinc, and uranium in the U.S., tin in Brazil, and copper in Mexico.

- **Weather satellites** record cloud patterns and movements, which help to predict storms. They also measure temperature, air pressure, rainfall, and snow depth.

- **Satellites can track** large fires, wildlife outfitted with radio transmitters, glaciers calving into icebergs, and changes in the size of the "hole" (thinning) in the ozone layer.

The Ages of the Earth

- **The Earth formed** 4,570 million years ago (mya), but the first animals with shells and bones appeared less than 600 mya. With the help of these fossils, geologists have learned about the Earth's history since then. We know very little about the 4,000 million years before, known as the Precambrian Eon.

- **Geologists divide** the Earth's history into time periods. The longest are eons, thousands of millions of years long. The shortest are chrons, a few thousand years long. In between come eras, periods, epochs, and ages.

- **The years since the Precambrian Eon** are split into three eras: Palaeozoic, Mesozoic, and Cenozoic.

- **Different plants and animals** lived at different times, so geologists can tell from the fossils in rocks how long ago the rocks formed. Using fossils, they have divided the Earth's history since Precambrian Time into 12 periods.

- **Layers of rock form** on top of each other, so the oldest rocks are usually at the bottom, unless they have been disturbed. The order of layers from top to bottom is known as the geological column.

- **By looking for certain fossils** geologists can tell if one layer of rock is older than another, and place it in the geological column.

- **Fossils can show** only if a rock is older or younger than another, they cannot give a date in years. Also many rocks, such as igneous, contain no fossils. To give an absolute date, geologists may use radiocarbon and other similar radio-dating methods.

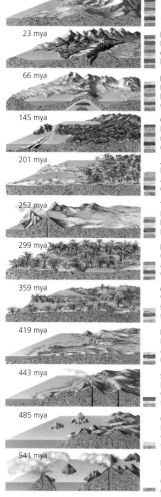

2.6 mya

Quaternary Period
Many mammals die out in ice ages; humans evolve

23 mya

Neogene Period
Modern kinds of carnivore mammals, like cats and wolves

66 mya

Paleogene Period
First large mammals; birds flourish; widespread grasslands

145 mya

Cretaceous Period
First flowering plants; non-bird dinosaurs die out

201 mya

Jurassic Period
Dinosaurs widespread; Archaeopteryx is the earliest known bird

252 mya

Triassic Period
First mammals; seed-bearing plants spread; Europe is in the tropics

299 mya

Permian Period
Conifers replace ferns as big trees; deserts are widespread

359 mya

Carboniferous Period
Vast warm swamps of fern forests form coal; first reptiles

419 mya

Devonian Period
First insects and amphibians; ferns and mosses as big as trees

443 mya

Silurian Period
First land plants; fish with jaws and freshwater fish

485 mya

Ordovician Period
Early fishlike vertebrates appear; the Sahara is glaciated

541 mya

Precambrian Eon
The first life forms (bacteria) appear, and give the air oxygen

◀ This sequence shows the main geological periods and events in Earth's history, from the most recent (top) to the most ancient.

● **Radio-dating** allows the oldest rocks on Earth to be dated. After certain substances, such as uranium and rubidium, form in rocks, their atoms slowly break down and they give off radioactivity. By assessing how many atoms have changed, geologists work out the rock's age.

● **Breaks in the sequence** of the geological column, called unconformities, can help to build up a picture of an area's geological history.

23

Ice ages

- **Ice ages** are periods lasting millions of years when the Earth is so cold that the polar ice caps grow huge. There are various theories about why they occur.

- **There have been five ice ages** in the last 1,000 million years, including one that lasted 100 million years.

- **The most recent ice age**—called the Pleistocene Ice Age—began about 2.5 mya.

- **In an ice age**, the climate varies between cold spells called glacials and warm spells called interglacials.

- **There have been** about 15 glacials and interglacials in the last 2.5 million years of the Pleistocene Ice Age.

▼ *California may have looked something like this 18,000 years ago when it was on the fringes of an ice sheet.*

- **The last glacial** peaked about 25,000–18,000 years ago and ended 12,000–11,000 years ago.

- **Ice covered 40 percent** of the world 18,000 years ago.

- **Glaciers spread** over much of Europe and North America 18,000 years ago. Ice caps grew in Tasmania and New Zealand.

- **About 18,000 years ago** there were glaciers in Hawaii and Australia.

DID YOU KNOW?

Where Washington D.C. and London are today, the ice was almost one mile thick 18,000 years ago.

Between ice ages

- **Warmer periods** in Earth's history are called interglacials, hyperthermals, or climate optima.

- **Fluctuations in global temperatures** are caused in part by the cyclical changes in the Earth's orbit around the Sun, its tilt angle toward the Sun, and the wobble of its axis.

▼ *During the Late Jurassic Period, 150 mya, great dinosaurs such as* Brachiosaurus *feasted on huge plants growing in the warm, moist climate.*

- **Clues to past climates** include tree rings in fossilized timber, ice cores from ancient glaciers and ice sheets, ocean sediments, coral reefs, and fossils in sedimentary rocks.

- **The highest global temperatures** over the last 200 million years were during the Cretaceous Thermal Optimum about 100 mya.

- **During this time** there was no ice at the Poles—only deciduous forest and tropical animals. It was 10.8–14.4°F warmer than today and carbon dioxide levels were five times higher.

- **The Paleocene-Eocene** Thermal Maximum (PETM) started around 55 million years ago and lasted about 170,000 years. The temperature rose by 10.8°C in 20,000 years.

- **The PETM** was probably caused by the release of frozen methane, a greenhouse gas, trapped at the bottom of the ocean.

- **In the last 2.6 million years**—the Pleistocene—there have been numerous interglacials, at roughly 40,000–100,000 year intervals, when tundra vegetation and ice sheets retreated toward the poles and were replaced by forest.

- **During the Pleistocene Eemian interglacial**, 130,000–114,000 years ago, sea levels were 26 ft higher than today and ocean water temperature was 3.6°F higher.

- **The Bølling-Allerød interstadial** near the end of the Pleistocene Period, 14,670 years ago, was a short warm period when sea levels rose by 328 ft.

- **The Earth is presently** in an interglacial period, the Holocene, when temperatures were quite high 9,000–5,000 years ago.

Fossils

- **Fossils are the remains** of living things that have been preserved for millions of years, usually in stone.

- **Most fossils** are the remains of living things such as bones, shells, claws, teeth, leaves, bark, cones, and seeds.

- **Trace fossils** are fossils of signs left behind by creatures, such as footprints and scratch marks.

- **Paleontologists** (scientists who study fossils) can tell the age of a fossil from the rock layer in which it is found. They can also measure how the rock has changed radioactively since it was formed (radio-dating).

- **The oldest fossils** are called stromatolites. They are fossils of big colonies of microscopic bacteria over 3,500 million years old.

- **The biggest fossils** are conyphytons—2,000-million-year-old stromatolites over 328 ft high.

◄ Transformed into hard, new minerals, remains of living things can be preserved for hundreds of millions of years, although fossils as good as these ancient fish are rare.

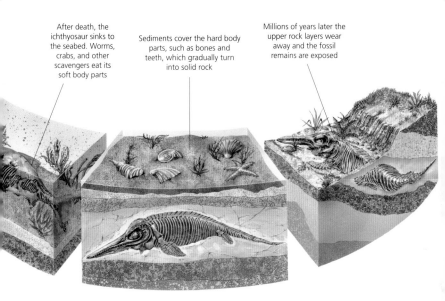

After death, the ichthyosaur sinks to the seabed. Worms, crabs, and other scavengers eat its soft body parts

Sediments cover the hard body parts, such as bones and teeth, which gradually turn into solid rock

Millions of years later the upper rock layers wear away and the fossil remains are exposed

▲ *Ichthyosaurs were giant marine reptiles that lived at the time of the dinosaurs. This sequence shows how one could have been fossilized.*

- **Not all fossils are stone**. Mammoths have been preserved by being frozen in the permafrost of Siberia.

- **Insects have been preserved** in amber, the solidified resin of ancient trees.

- **Certain widespread**, short-lived fossils are very useful for dating rock layers. These are known as index fossils.

- **Index fossils include** shellfish such as brachiopods, belemnites, and ammonites, also trilobites, graptolites, and crinoids.

The Earth's chemistry

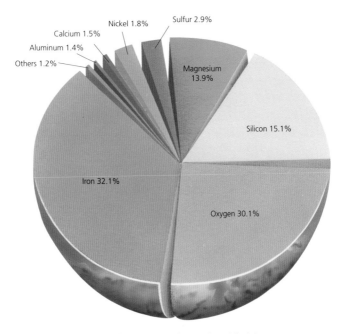

Nickel 1.8%

Sulfur 2.9%

Calcium 1.5%

Aluminum 1.4%

Others 1.2%

Magnesium 13.9%

Silicon 15.1%

Iron 32.1%

Oxygen 30.1%

▲ *This diagram shows the percentages by mass (weight) of the chemical elements that make up the Earth.*

● **The bulk of the Earth** is made from iron, oxygen, silicon, and magnesium.

● **More than 80 chemical elements** occur naturally in the Earth and its atmosphere.

- **The crust** is made mostly from oxygen and silicon, with aluminum, iron, calcium, magnesium, sodium, potassium, titanium, and traces of 64 other elements.

- **The upper mantle** is iron and magnesium silicates. The lower is silicon and magnesium sulfides and oxides.

- **The core is mostly iron**, with a small amount of nickel and traces of sulfur, carbon, oxygen, and potassium.

- **Evidence for the Earth's chemistry** comes from analyzing its density with the help of earthquake waves, and from studying stars, meteorites, and other planets.

- **When the Earth was still semi-molten**, dense elements such as iron sank to form the core. Lighter elements such as oxygen floated up to form the crust.

- **Some heavy elements**, such as uranium, ended up in the crust because they make compounds easily with oxygen and silicon.

- **Large amounts** of elements that combine easily with sulfur, such as zinc and lead, spread through the mantle.

DID YOU KNOW?
Earth's rarest naturally occurring chemical element is astatine, with a total weight of one gram.

- **Elements that combine** easily with iron, such as gold and nickel, sank to the core.

Minerals

- **Minerals are the natural chemicals** from which rocks are made. Some rocks are made from crystals of just one mineral, but many consist of several.

- **Most minerals** are made up of two or more chemical elements, but a few, such as gold and copper, consist of just one element.

- **There are over 2,000 minerals**, and around 30 of these are very common. Less common minerals are present in rocks in minute traces, but they may become concentrated in certain places by geological processes.

- **The Earth's surface** contains an enormous wealth of mineral resources, from clay for bricks to precious gems.

▼ Bulk materials such as cement, gravel, and clay are taken from the ground in huge quantities for building.

- **Silicates make up** the largest group of minerals. They form when metals join with oxygen and silicon. The most common silicates are quartz and feldspar, which are rock-forming minerals.

- **Other common minerals** are oxides such as hematite and cuprite, sulfates such as gypsum and barite, sulfides such as galena and pyrite, and carbonates such as calcite and aragonite.

- **Fossil fuels**—oil, coal, and natural gas—formed from the remains of plants and animals that lived millions of years ago. The remains changed into fuel by intense heat and pressure. Coal formed from plants that grew in huge, warm swamps. Oil and natural gas form from the remains of tiny marine plants and animals.

- **Ores are the minerals** from which metals are extracted. Bauxite is the ore for aluminum, chalcopyrite for copper, galena for lead, hematite for iron, and sphalerite for zinc.

- **Some minerals**, like amber and opal, do not have a microcrystal structure and are called mineraloids.

Rocks

- **There are three main kinds of rock**: igneous, sedimentary, and metamorphic.

- **The oldest known rocks** are in Canada—four-billion-year-old Acasta gneiss, and maybe even older is Nuvvuagittuq greenstone near Hudson Bay.

- **The youngest rocks** form today as molten lava erupts from volcanoes, cools, and solidifies into solid rock.

- **The general scientific study** of rocks, their structures, and how they change is known as geology.

- **Petrology** is a specialism within geology that studies the detailed composition and microstructure of rocks, their origins, and their fates.

- **Mineralogy** is the scientific study of the minerals that make up rocks, their origins, and how they change or react when rocks change form.

- **Geology** and its allied sciences are vital when surveying or prospecting for mineral wealth such as coal, petroleum oil, natural gas, and ores such as those containing metals or sulfur.

- **One of the heaviest** or densest rocks is peridotite, with a density of up to 0.1lb per cu in.

Erosion and transport

Laying down of sediment

Sediment is buried and compacted

▶ Rocks are continually recycled. They are broken down by weathering and erosion into sediments such as sand and mud. These settle on the beds of seas, lakes, and rivers, where they harden to form new rock, in the ongoing process called the rock cycle.

● **This compares** with the Earth's most dense naturally occurring element or pure chemical substance, the metal osmium, with a density of 0.8 lb per cu in.

● **Other very dense** or heavy rocks include basalt, diabase, and diorite.

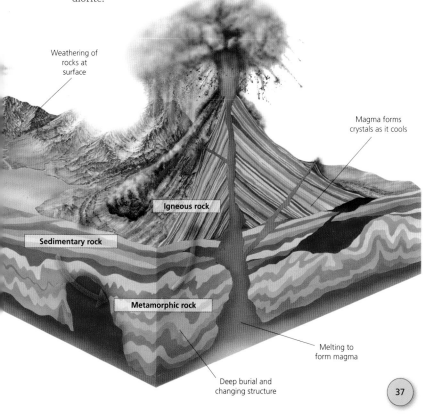

Weathering of rocks at surface

Magma forms crystals as it cools

Igneous rock

Sedimentary rock

Metamorphic rock

Melting to form magma

Deep burial and changing structure

Igneous rocks

- **Igneous rocks are formed** when hot liquid magma or lava is cooled and solidifies.

- **If cooling occurs below** the Earth's surface, usually of magma, intrusive or plutonic igneous rock such as granite is formed.

- **If cooling occurs above** the Earth's surface, when lava erupts from a volcano, for example, extrusive or volcanic igneous rock such as basalt is formed.

- **If the rock cools rapidly**, as surface lava does, the crystals formed are very small or non-existent. Such rocks are called aphanitic.

- **If the rock cools slowly**, while underground for example, the crystals are much bigger, producing a coarse-grained rock, called phaneritic.

- **The upper parts** of the Earth's crust, to a depth of about 10 mi, consist of 90–95 percent igneous rocks.

- **Classification of igneous rocks** depends on how they were formed, the texture of the crystals, but most importantly the minerals and chemicals they contain.

- **Over 700 different types** of igneous rocks have been described and classified.

- **Granite is classed** as a felsic igneous rock—it contains lots of quartz and feldspar and is light-colored.

- **Basalt is classified** as a mafic igneous rock—it contains less quartz and less feldspar than granite and is usually dark-colored.

▼ *Peaks of the Stawamus Chief Provincial Park in British Columbia, Canada, are made of the very hard igneous rock granite. It formed about 100 million years ago and has been eroded by glaciers.*

Metamorphic rocks

- **"Metamorphism"** means "to change form." Metamorphic rocks have been changed by extreme heat and pressure but without melting.

- **The rocks may be formed deep** in the Earth, or nearer the surface by volcanoes or tectonic movements, and are classified by their texture and chemical composition.

- **They may originally** have been igneous or sedimentary or even metamorphic rocks, and may have been changed physically or chemically or both.

- **The physical and chemical** make-up depends on the type of original rock and the temperatures and pressures it has been subjected to.

▼ Marble is a pale metamorphic rock containing mostly carbonate minerals. Its delicate colors and swirls make quarrying it big business.

▲ *The beautiful Taj Mahal tomb in Agra, India, was constructed in the 1650s from gleaming white marble.*

- **Recrystallization often occurs.** This is when crystals are changed to larger forms, for example, fine-grained limestone turns to large-crystal marble.

- **The physical structure** often reflects the type of forces that have pulled and pushed the rock as it reformed, with swirls and folds.

- **Contact metamorphism** occurs when hot liquid magma flows next to the rock. Many mineral ores form like this.

- **Regional metamorphism** occurs at great depths or under tectonic movements. These rocks are often foliated or layered, such as slate.

- **Coarse-grained rock**, such as schist, and very coarse-grained rock, such as gneiss, were probably formed by uniform pressure.

- **Temperatures** and pressures can be so great that chemicals can move between rock types while still remaining solid.

41

Sedimentary rocks

- **Sedimentary rocks** are formed from layers of sediments, such as sand, mud, and silt. These are created by eroded particles of rock, which are carried and then laid down in seas, lakes, and rivers, or blown by winds on land.

- **The sediments are subjected** to extreme pressure by the weight of more sediment deposited on top, and also chemical alteration, and this changes them to rock.

- **Sedimentary rocks** are only thinly deposited over the Earth's crust and are thought to form only about 8 percent of it.

- **These rocks form strata** or beds that reflect the layers and order in which they were deposited.

- **Sedimentary rocks show** the Earth's history by preserving a physical record of such features as riverbeds, beaches, sand dunes, or life-forms.

- **Clastic sedimentary rocks**, such as sandstones, are formed from sediments left beneath rivers and seas.

- **Biochemical sedimentary rocks** are formed from the shells and other remains of living things and include limestone and coal.

- **Chemical sedimentary rocks** are formed when dissolved chemicals become supersaturated and precipitate or crystallize out of solution; rock salt or halite is an example.

- **Sedimentary rocks often contain fossils**, the remains or traces of living things that become permineralized and so preserved.

▶ *The sedimentary rocks in Canyonlands National Park in Utah, U.S.A. include sandstones, limestones, and shales. They have been eroded into flat-topped, steep-sided shapes called mesas.*

Earth's interior

- **The Earth has three main layers** inside. Outermost is the crust, a thin, hard shell of rock, which is a few dozen miles thick.

- **The crust's thickness** in relation to the size of the whole Earth is about the same as the skin on an apple.

- **Under the crust**, there is a deep layer of hot, soft rock called the mantle.

- **Beneath the mantle** is a core of hot iron and nickel that is mostly semi-molten or melted.

- **The inner core** contains 1.7 percent of the Earth's mass, the outer core 30.8 percent, the core–mantle boundary 3 percent, the lower mantle 49 percent, the upper mantle 15 percent, the ocean crust 0.099 percent, and the continental crust 0.374 percent.

- **Satellite measurements** are so accurate they can detect slight lumps and dents in the Earth's surface. These indicate where gravity is stronger or weaker because of differences in rock density. Variations in gravity reveal features such as mantle plumes (upwellings of unusually hot rock within the mantle).

- **Our knowledge** of the Earth's interior comes mainly from studying how earthquake waves vibrate through the Earth.

- **Analysis of how** earthquake waves are deflected reveals where different materials occur in the interior. S (secondary) waves pass only through the mantle. P (primary) waves pass through the core as well.

- **As P waves pass through** the core they are deflected, leaving a shadow zone where no waves reach the far side of the Earth.

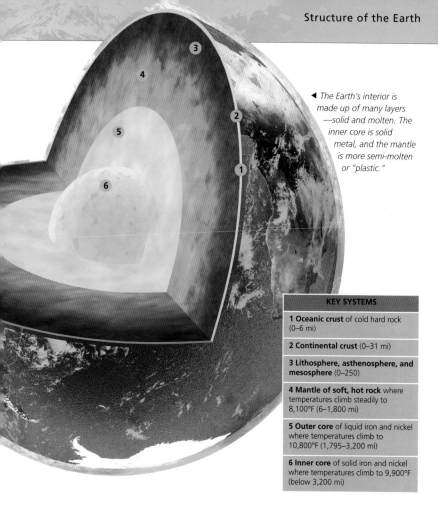

◀ The Earth's interior is
made up of many layers
—solid and molten. The
inner core is solid
metal, and the mantle
is more semi-molten
or "plastic."

KEY SYSTEMS
1 Oceanic crust of cold hard rock (0–6 mi)
2 Continental crust (0–31 mi)
3 Lithosphere, asthenosphere, and mesosphere (0–250)
4 Mantle of soft, hot rock where temperatures climb steadily to 8,100°F (6–1,800 mi)
5 Outer core of liquid iron and nickel where temperatures climb to 10,800°F (1,795–3,200 mi)
6 Inner core of solid iron and nickel where temperatures climb to 9,900°F (below 3,200 mi)

● **The speed at which** earthquake waves travel reveals how dense the rocks are. Cold, hard rock transmits waves more quickly than hot, soft rock.

Earth's core

- **The core**, the deepest layer of rock that makes up the Earth, has two layers—the inner core and the outer core.

- **The inner core** is thought to be solid while the outer core is a viscous liquid, like treacle.

- **These layers** were discovered by measuring how long seismic waves, caused by earthquakes, take to travel through the Earth.

- **The inner core starts** at the center of the Earth and finishes about 3,100–3,700 mi beneath the surface.

- **The outer core starts** at this depth and finishes 1,800–3,100 mi beneath the surface.

- **When the young Earth** was still molten the heavier elements sank to the middle, so the inner core is thought to be made of iron and nickel.

- **The density of the inner core** is about 12–13 tons per cu m, that of the outer core is 10–12 tons per cu m.

- **Studies show** that the inner core may be made up of very large crystals or even a single giant crystal.

- **The outer core** may also be made up of iron and nickel but probably contains other, lighter elements as well.

- **The Earth's magnetic field** is probably caused by the liquid metals moving in the outer core as the Earth spins, causing a dynamo effect.

DID YOU KNOW?

In the search for oil, exploratory drilling ships make boreholes that are more than 32,000 ft deep.

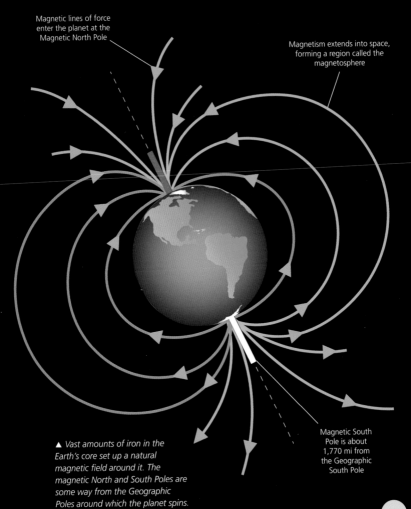

Magnetic lines of force enter the planet at the Magnetic North Pole

Magnetism extends into space, forming a region called the magnetosphere

Magnetic South Pole is about 1,770 mi from the Geographic South Pole

▲ *Vast amounts of iron in the Earth's core set up a natural magnetic field around it. The magnetic North and South Poles are some way from the Geographic Poles around which the planet spins.*

The mantle

- **The mantle**, the layer of rock between the Earth's core and its crust, starts about 21–37 mi beneath the surface and is about 1,800 mi thick.

- **This is the thickest layer** within the Earth and makes up about 84 percent of the planet's total volume and two-thirds of its mass.

- **Mantle rock** is mainly made up of silicates (silicon and oxygen) rich in iron and magnesium.

- **The boundary between** the mantle and the overlying crust is called the Moho Discontinuity.

- **There are three layers** to the mantle: the upper mantle, the transition zone, and the lower mantle.

- **The temperature of the mantle** varies between 900–1,650°F in its upper regions to 8,100°F near its base.

- **Despite these temperatures** the high pressures increase the melting point of the rock and it remains in a solid/plastic state that flows very slowly.

- **Heat creeps up from great depth**, around the outer mantle and back down again, a circulation called convection currents.

- **These convection currents** bring rock to the surface at spreading rifts and carry them down again at subduction zones.

- **The mantle rock** is thought to contain as much water again as is contained in all the oceans.

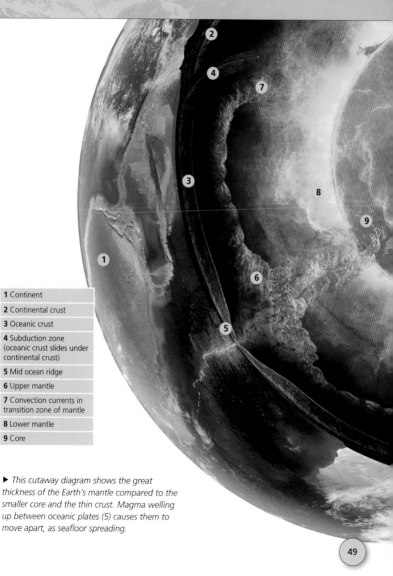

1 Continent

2 Continental crust

3 Oceanic crust

4 Subduction zone
(oceanic crust slides under
continental crust)

5 Mid ocean ridge

6 Upper mantle

7 Convection currents in
transition zone of mantle

8 Lower mantle

9 Core

▶ This cutaway diagram shows the great
thickness of the Earth's mantle compared to the
smaller core and the thin crust. Magma welling
up between oceanic plates (5) causes them to
move apart, as seafloor spreading.

The crust

- **The Earth's crust** is its hard outer shell.

- **The crust is a thin layer** of dense, solid rock that floats on the mantle. It is made mainly of silicate minerals (minerals made of silicon and oxygen) such as quartz.

- **There are two kinds of crust—** oceanic and continental.

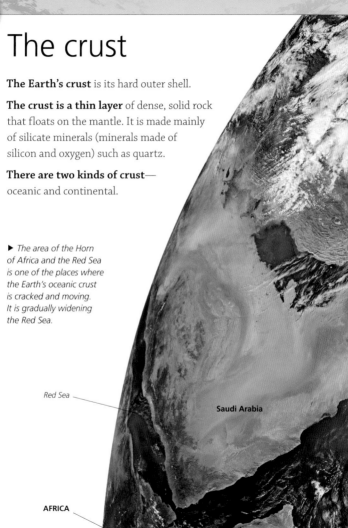

▶ The area of the Horn of Africa and the Red Sea is one of the places where the Earth's oceanic crust is cracked and moving. It is gradually widening the Red Sea.

Red Sea

Saudi Arabia

AFRICA

- **Oceanic crust** is the crust beneath the oceans. It is much thinner—just 4 mi thick on average. It is also young, with none being over 200 million years old.

- **The crust beneath** the continents is called continental crust. It is up to 50 mi thick and mostly old.

DID YOU KNOW?
The crust forms less than one percent of Earth's volume and is less than 0.5 percent of its mass.

- **Continental crust** is mostly crystalline "basement" rock up to 3,800 million years old. Some geologists think at least half of this rock is over 2,500 million years old.

- **About 0.2 cu mi** of new continental crust is probably being created each year.

- **The "basement" rock** has two main layers: an upper half of silica-rich rock, such as granite, schist, and gneiss, and a lower half of volcanic rock, such as basalt, which has less silica. Oceanic crust is mostly basalt.

- **Continental crust** is created in the volcanic arcs above subduction zones. Molten rock from the subducted plate oozes to the surface over a period of a few hundred thousand years.

- **Almost one half** (47 percent) of the crust by mass is the chemical element oxygen, and more than one-quarter (28 percent) is silicon.

Lithosphere

- **The outer, rigid layer** of the Earth is called the lithosphere. It consists of the crust and the top of the mantle. It is about 62 mi thick.

- **The lithosphere was discovered** by seismology—the study of the patterns of vibrations from earthquakes.

- **It is thickest**—around 75 mi—below the continents.

- **The lithosphere is only** a few miles thick below the middle of the oceans. Here, the mantle's temperature just below the surface is 2,370°F.

- **Fast earthquake waves** show that the top of the mantle is as rigid as the crust, although it is chemically different.

- **Lithosphere** means "ball of stone."

- **The lithosphere is broken up** into about 20 slabs called tectonic plates. The continents sit on top of the continental plates.

- **Temperatures increase** by 63°F for every 3,280 ft you move down through the lithosphere.

- **Below the lithosphere**, in the Earth's mantle, is the hot, soft rock of the asthenosphere.

- **The boundary between** the lithosphere and the asthenosphere occurs at the point where temperatures climb above 2,300°F.

◄ The hard, rocky surface
of the Earth is made up of
20 or so strong, rigid
plates of lithosphere.

Continental drift

1 It is thought that Earth's landmass was once gathered into a vast single continent called Rodinia, which began to break apart around 750 million years ago.

2 By about 220 million years ago, all the continents had moved together again, forming the supercontinent of Pangaea. A single enormous ocean, known as Panthalassa, surrounded the landmass.

- **Continental drift** is the slow movement of the continents around the world.

- **About 220 million years ago** all the continents were joined together in one supercontinent, which geologists call Pangaea.

- **Pangaea** began to break up about 190 million years ago. The fragments slowly drifted apart to form the continents we know today.

- **South America** used to be joined to Africa. North America used to be joined to Europe.

- **The first hint that the continents were once joined** was the discovery made by German explorer Alexander von Humboldt (1769–1859) that rocks in Brazil (South America) and the Congo (Africa) are very similar.

- **When German meteorologist** Alfred Wegener (1880–1930) first suggested the idea of continental drift in 1923, many scientists were sceptical.

● **Evidence of continental drift** has come from similar ancient fossils found on separate continents. These include the *Glossopteris* fern found in Australia and India (Asia), diadectid reptiles found in Europe and North America, and *Lystrosaurus*—a reptilelike creature from 200 million years ago—found in Africa, India and China (Asia), and Antarctica.

● **Satellites provide** incredibly accurate ways of measuring and can record the slow movement of the continents. The main method is satellite laser ranging (SLR), where laser beams are bounced off a satellite from ground stations on each continent. Other methods include the Global Positioning System (GPS) and Very Long Baseline Interferometry (VLBI).

● **Rates of continental drift** vary. India drifted north into Asia relatively quickly. South America is moving 8 in further from Africa every year. On average, continents move at the same rate as fingernails grow.

3 Today, North America is still moving away from Europe and closer to Asia.

4 The next supercontinent—50–200 million years in the future—is predicted to be Amasia, forming from the meeting of the Americas and Asia.

◀ It is hard to believe that the continents move—over tens of millions of years they move huge distances. The drifting of the continents has changed the map of the world very slowly over the past 200 million years, and will continue to do so.

55

Tectonic plates

- **The Earth's surface** is divided into thick slabs called tectonic plates. Each plate is a fragment of the Earth's rigid outer layer, or lithosphere.

- **There are about** ten larger plates, and dozens of smaller ones.

- **The biggest** is the Pacific plate, which lies beneath the whole of the Pacific Ocean.

- **Tectonic plates are moving** all the time, by an average of about 4 in per year. Over millions of years they move vast distances. Some have moved halfway around the globe.

- **Continents are embedded** within most of the plates and move with them.

- **The Pacific plate** is the only large plate with no part of a continent situated on it.

- **The movement of tectonic plates** accounts for many geological events, including the patterns of volcanic and earthquake activity around the world.

- **There are three kinds** of boundary between plates: convergent, divergent, and transform.

- **Tectonic plates** are probably driven by convection currents of molten rock that circulate within the Earth's mantle.

- **The lithosphere** was too thin for tectonic plates to form until 500 million years ago.

DID YOU KNOW?

Using GPS (Global Positioning System) satellites, the minute movements of the plates can be tracked day by day.

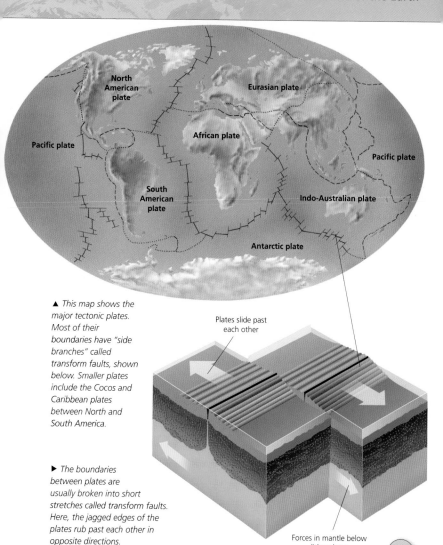

North American plate

Eurasian plate

African plate

Pacific plate

Pacific plate

South American plate

Indo-Australian plate

Antarctic plate

▲ This map shows the major tectonic plates. Most of their boundaries have "side branches" called transform faults, shown below. Smaller plates include the Cocos and Caribbean plates between North and South America.

Plates slide past each other

▶ The boundaries between plates are usually broken into short stretches called transform faults. Here, the jagged edges of the plates rub past each other in opposite directions.

Forces in mantle below lithosphere

57

Converging plates

- **In many places** around the world, the tectonic plates that make up the Earth's crust are slowly crunching together with huge force.

- **The Atlantic Ocean** is getting wider, pushing apart the Americas from Europe and Africa. But the Earth is not getting any bigger.

- **As the west edges** of the American plates crash into the Pacific ocean plate, the thinner, denser ocean plate is driven down into the Earth's hot mantle and melted.

- **The process of driving** an ocean plate down into the Earth's interior is called subduction.

- **Subduction creates** deep ocean trenches typically 3–4.5 mi deep. One of these, the Mariana Trench, is 7 mi at its deepest point.

▼ *This cross-section through the top 600 mi or so of the Earth's surface shows a subduction zone—where an ocean plate is bent down beneath a continental plate.*

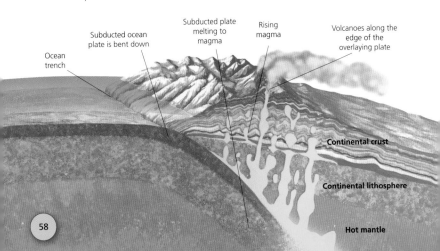

Subducted plate melting to magma

Rising magma

Volcanoes along the edge of the overlaying plate

Subducted ocean plate is bent down

Ocean trench

Continental crust

Continental lithosphere

Hot mantle

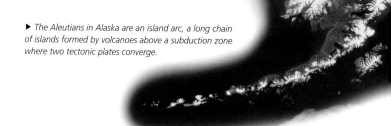

▶ *The Aleutians in Alaska are an island arc, a long chain of islands formed by volcanoes above a subduction zone where two tectonic plates converge.*

● **As an oceanic plate bends down** into the Earth's mantle, it cracks. The movement of these cracks sets off earthquakes originating up to 435 mi below the surface. These earthquake zones are called Benioff–Wadati zones after two experts, Hugo Benioff (1899–1968) of the U.S.A and Kiyoo Wadati (1902–1995) of Japan, who discovered them separately in the 1950s.

● **An oceanic plate melts** as it slides down, and creates blobs of magma. This magma floats up toward the surface, pushing its way through weak crust to create a line of volcanoes along the edge of the continental plate.

● **If volcanoes in subduction zones emerge** in the sea, they form a curving line of volcanic islands called an island arc. Beyond this arc is the back-arc basin, an area of shallow sea that slowly fills up with sediments.

● **As a subducting plate sinks**, the continental plate scrapes sediments off the ocean plate and piles them in a great wedge. Between this wedge and the island arc there may be a fore-arc basin, which is a shallow sea that slowly fills with sediment.

● **Where two continental plates collide**, the plates split into two layers: a lower layer of dense upper-mantle rock and an upper layer of lighter crustal rock, which is too buoyant to be subducted. As the mantle rock goes down, the crustal rock peels off and crumples against the other to form fold mountains.

Diverging plates

- **Deep down on the ocean floor**, some of the tectonic plates of the Earth's crust are slowly pushing apart. New molten rock wells up from the mantle into the gap between them and solidifies onto their edges. As plates are destroyed at subduction zones, newly made plate spreads the ocean floor wider, known as seafloor spreading.

- **The spreading or divergence** of the ocean floor centers on long ridges along the middle of some oceans, called mid-ocean ridges. Some of these ridges are joined, forming the world's longest mountain range, which winds over 37,300 mi beneath the oceans.

- **The Mid Atlantic Ridge** stretches through the Atlantic Ocean from the North Pole to the South Pole. The East Pacific Rise mid-ocean ridge winds under the Pacific Ocean from Mexico to Antarctica.

- **Along the middle** of a mid-ocean ridge is a deep canyon. This is where molten rock from the mantle wells up through the seabed.

- **Mid-ocean ridges** are broken by the curve of the Earth's surface into short, stepped sections. Each section is marked off by a long sideways crack called a transform fault. As the seafloor spreads out from a ridge, the sides of the fault rub together, setting off earthquakes.

- **As molten rock wells up** from a ridge, cools, and hardens, its magnetic material solidifies in a certain way to line up with the Earth's magnetic field. Because the field reverses every now and then, bands of material harden with magnetism in alternate directions. This means that scientists can see how the seafloor has spread in the past.

DID YOU KNOW?

About 2.4 cubic mi of new crust is created at mid-ocean ridges every year.

● **Rates of seafloor spreading** vary from 0.4–8 in a year. Slow-spreading ridges, such as the Mid Atlantic Ridge, are higher, with sea mounts often topping them. Fast-spreading ridges, such as the East Pacific Rise, are lower, and magma oozes from these like fissure volcanoes on the surface.

● **Hot magma bubbling up** through a mid-ocean ridge emerges as hot lava. As it comes into contact with the cold sea water it solidifies into blobs called pillow lava.

● **Mid-ocean ridges may begin** where mantle plumes rise through the mantle and melt through the seabed. Plumes may also melt through continents to form Y-shaped cracks, which begin as rift valleys and then widen into new oceans.

▼ This cross-section of the top 30 mi or so of the Earth's surface shows how the seafloor is spreading away from the mid-ocean ridge.

Ridges are lower and older the further away from the center

Mantle

Ocean plate

Mid-ocean ridge

Magma erupts through the gap as lava solidifies into new seafloor

Transform fault

Faults

- **A fault is a fracture** in rock along which large blocks of rock have slipped past each other.

- **Faults usually occur** in fault zones, which are often along the boundaries between tectonic plates. Faults are typically caused by earthquakes.

- **Single earthquakes** rarely move blocks more than a few inches. Repeated small earthquakes can shift blocks hundreds of miles.

- **Compression faults** are caused by rocks being squeezed together, perhaps by converging plates.

- **Tension faults** are caused by rocks being pulled together, perhaps by diverging plates.

▶ The San Andreas Fault in California is the most famous example of a transcurrent fault.

- **Normal, or dip-slip, faults** are tension faults where the rock fractures and slips straight down.

- **A wrench, or tear, fault** occurs when plates slide past each other and make blocks slip horizontally.

- **Large wrench faults**, such as the San Andreas Fault in California, U.S.A, are called transcurrent faults.

- **Rift valleys are huge**, trough-shaped valleys created by faulting, such as Africa's Great Rift Valley. The valley floor is a depressed block called a graben. Some geologists think they are caused by tension, others by compression.

- **Horst blocks** are blocks of rock thrown up between normal faults, often creating a high plateau.

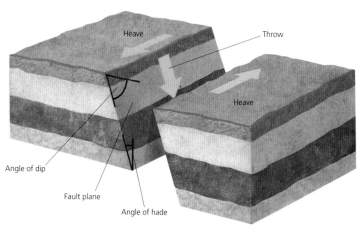

▲ Geologists who study faults describe their movement using the terms illustrated here.

Folds

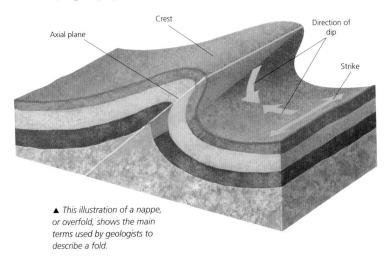

Axial plane

Crest

Direction of dip

Strike

▲ *This illustration of a nappe, or overfold, shows the main terms used by geologists to describe a fold.*

● **Rocks usually form** in flat layers called strata. Tectonic plates can collide with such force that they crumple up these strata.

● **Sometimes the folds** are just tiny wrinkles a few inches long. Sometimes they are huge, with hundreds of miles between crests (the highest points on a fold).

● **The shape of a fold** depends on the force that is squeezing it and on the resistance of the rock.

● **The slope of a fold** is called the dip. The direction of the dip is the direction in which it is sloping.

- **The strike of the fold** is at right angles to the dip. It is the horizontal alignment of the fold.

- **Some folds** turn right over on themselves to form upturned folds called nappes.

- **As nappes fold on top** of other nappes, the crumpled strata may pile up into mountains.

- **A downfold** is called a syncline. An upfolded arch of strata is called an anticline.

- **The axial plane** of a fold divides the fold into two halves.

▼ The bent and folded layers of rock are clearly visible in this cliff in the Ugab Valley, Damaraland, Namibia.

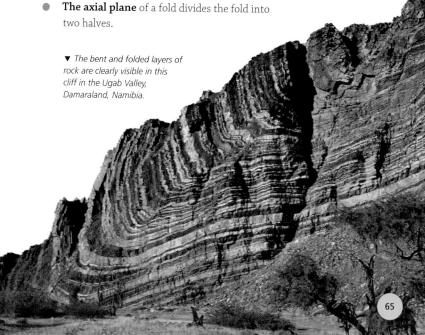

65

High mountains

▼ High peaks such as Mount Everest are jagged because massive folding fractures the rock and makes it vulnerable to the sharp frosts and erosion.

- **A few high mountains**, such as Africa's Kilimanjaro, are lone volcanoes that are built up by eruptions.

- **Some volcanic mountains**, such as Japan's Mount Fuji, are in chains in volcanic arcs.

- **Most high mountains** are part of great mountain ranges stretching over hundreds of miles.

- **Some mountain ranges** are huge slabs of rock called fault blocks. They were forced up by quakes.

- **Other mountain ranges**, such as the Himalayas in Asia and the Andes in South America, are fold mountains.

- **The height of mountains** used to be measured on the ground, using levels and sighting devices to measure angles. Now, mountains are measured accurately using satellite techniques.

- **Satellite measurements** taken in 1999 raised the height of the world's highest peak, Mount Everest in the Himalayas, Nepal, from 29,028 ft to 29,035 ft.

- **Everest's height** was then revised again back to 29,028 ft.

- **Temperatures drop** 1.08°F for every 300 ft you climb, so mountain peaks are very cold and often covered in snow.

- **Air is thinner** high up on mountains, so the air pressure is lower.

- **There is also less oxygen** in the thinner air. Animals are specially adapted with larger lungs and extra red blood cells.

- **Climbers may need** oxygen masks to breathe, especially above 26,000 ft.

Mountain ranges

- **Great mountain ranges**, such as the Andes in South America, usually lie along the edges of continents.

- **Most mountain ranges** are formed by the folding of rock layers as tectonic plates move together.

- **High mountain ranges** are geologically young because they are not yet worn down.

- **Mountain building** is very slow because rocks flow like thick treacle. Rock is pushed up like the bow wave in front of a boat as one tectonic plate pushes into another.

▼ Mountain ranges such as the Alps in Europe are thrown up by the crumpling of rock layers as the tectonic plates crunch together.

- **Satellite techniques** show that the central peaks of the Andes and the Himalayas are rising. The outer peaks are sinking as the rock flows slowly away from the "bow wave."

- **Mountain building is very active** during orogenic (mountain forming) phases, which last millions of years.

- **The process by which mountains form** makes the Earth's crust especially thick beneath them, giving mountains deep "roots."

- **As mountains are worn down**, their weight reduces and the "roots" float upward. This is called isostasy.

69

Volcanoes and earthquakes

Volcanoes

- **Volcanoes are places** on the Earth's crust where magma (hot, liquid rock from the Earth's interior) erupts or flows onto the surface.

- **The word "volcano"** comes from the name of Vulcano, a volcanic island in the Mediterranean. In ancient Roman mythology, Vulcan, the god of fire and blacksmith to the gods, was supposed to have forged his weapons in the fire beneath the mountain on Vulcano.

- **There are many types of volcano**. The most distinctive are cone-shaped composite volcanoes, which build up from alternating layers of ash and lava in successive eruptions.

- **Beneath a composite volcano** there is typically a large reservoir of magma called a magma chamber. Magma collects in the chamber before an eruption.

- **From the magma chamber** a narrow chimney, or vent, leads up to the surface. It erupts through the cone of debris at the peak, built up from previous eruptions.

- **When a volcano erupts**, the magma is driven up the vent by the gases within it. As the magma nears the surface, the pressure on it drops, which allows the gases dissolved in the magma to form bubbles. The expanding gases—mostly carbon dioxide and steam—push the molten rock upward and out of the vent.

▶ *This cutaway of an erupting volcano reveals the central vent through which red-hot molten rock moves up. In reality, the magma chamber is deep underground, between 1–6 mi below the surface.*

Old volcanic material from previous eruptions clogs up the vent. A new eruption shatters the plug into tiny pieces of ash and cinder, and blasts them high into the air

● **If the magma level** in the magma chamber drops, the top of the volcano's cone may collapse into it. This forms a crater called a caldera, which is Spanish for "boiling pot."

● **The world's largest caldera** is Toba on the island of Sumatra, Indonesia, which is 685 sq mi.

● **When a caldera subsides,** the whole cone may collapse into the old magma chamber. The caldera may fill with water to form a crater lake, such as Crater Lake in Oregon, U.S.A.

● **Not all of the magma gushes** up the central vent. Some exits through branching side vents, which often have their own small "parasitic" cones on the side of the main one.

Among the many kinds of ejected material, called tephra, are volcanic bombs that are fragments of the shattered volcanic plug flung out far and wide

Types of volcano

- **Each volcano** and eruption is slightly different.

- **Shield volcanoes** are shaped like curved shields or low domes. They erupt runny lava, which spreads over a wide area.

- **Fissure volcanoes** are places where floods of lava pour out of long cracks in the ground.

- **Composite volcanoes** are cone shaped. They build up in layers from a succession of explosive eruptions.

- **Cinder cones** are built up from ash, with little lava.

- **Strombolian eruptions** are moderately explosive eruptions of thick, sticky magma. They spit out sizzling blobs of lava called lava bombs.

- **Vulcanian eruptions** are explosive, erupting thick, sticky magma. The magma clogs the volcano's vent in between cannonlike blasts of clouds of ash and thick lava flows.

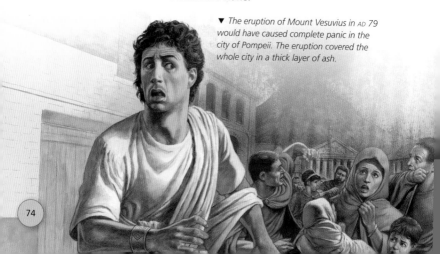

▼ The eruption of Mount Vesuvius in AD 79 would have caused complete panic in the city of Pompeii. The eruption covered the whole city in a thick layer of ash.

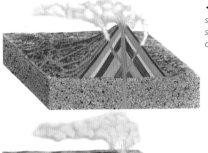

◀ *Thick magma creates explosive, cone-shaped volcanoes, made partly of ash, sometimes with a crater. Runnier magma creates flatter volcanoes that ooze lava.*

A composite volcano, or stratovolcano, has steep sides built up of layers of lava and ash

A shield volcano has a low, wide shape, with gently sloping sides

A caldera is a huge crater left after an old eruption. New cones often grow again inside

- **Peléean or Pelén** eruptions are the most violent, and eject glowing avalanches of ash and gas called *nuée ardentes* (French for "glowing clouds").

- **Plinian eruptions** are the most explosive type of eruption. They are named after the Roman writer and legal expert Pliny who witnessed the eruption of Vesuvius, in Italy, in AD 79.

- **During Plinian eruptions**, boiling gases blast clouds of ash and volcanic fragments many miles up into the atmosphere.

Volcanic eruptions

- **Volcanoes erupt** because of magma—the hot, liquid rock beneath the Earth's surface. Magma is less dense than the rock above, and so it "floats" or moves up toward the surface.

- **When magma is runny**, eruptions are "effusive," which means they ooze lava all the time.

- **When magma is thick and sticky**, eruptions are explosive. Magma clogs up the volcano's vent until enough pressure builds up to blast it out, like a popping cork.

- **An explosive eruption** blasts globs of hot magma, ash, cinder, gas, and steam high up into the air.

- **Shattered fragments** of the volcanic plug that are blasted out are called pyroclasts, from the ancient Greek for "fire broken."

- **Volcanoes usually erupt** again and again. The interval between eruptions, called the repose time, varies from a few minutes to thousands of years.

- **Magma near subduction zones** contains ten times more gas, so eruptions here are violent.

▶ *The mighty 1980 eruption of Mount St. Helens in Washington, U.S.A, measured 5 on the VEI.*

VE 8—Mega-colossal
Ash column height 25 km +
Volume erupted 1,000 km³

● **The gas inside magma** can expand hundreds of times in just a few seconds.

● **Volcanoes that erupt regularly** are known as active.

● **Volcanoes that are inactive** but could erupt in the future are dormant or "sleeping."

● **A volcano that is old and dead**, with no future eruptions possible, is called extinct.

VE 7—Super-colossal
Ash column height 25 km +
Volume erupted 100 km³

VE 6—Colossal
Ash column height 25 km +
Volume erupted 10 km³

▶ The Volcanic Explosivity Index (VEI) provides a measure of the power of eruptions. Each stage represents a ten-fold increase in explosivity.

VE 5—Paroxysmal
Ash column height 25 km +
Volume erupted 1 km³

VE 4—Cataclysmic
Ash column height 10–25 km
Volume erupted 100,000000 km³

VE 3—Severe
Ash column height 3–15 km
Volume erupted 10,000000 m³

VE 2—Explosive
Ash column height 1–5 km
Volume erupted 1,000,000 m³

VE 1—Gentle
Ash column height 100–1,000 m
Volume erupted 10,000 m³

Lava and ash

- **When a volcano erupts** it sends out a variety of hot materials, including lava, rock fragments, cinders, ash, and gases.

- **Lava is the name** for hot molten rock once it has been erupted. It is called magma while it is still underground.

▼ *Island volcanoes often ooze chunky lava, called "aa" lava, like this flow on the island of Réunion, in the Indian Ocean.*

- **Tephra is the material** blasted into the air by an eruption. It includes pyroclasts and volcanic bombs.

- **Pyroclasts are large fragments** of solidified volcanic rock that are thrown out by explosive volcanoes when the plug in the volcano's vent shatters. Pyroclasts are usually 1–3.2 ft across.

- **Large eruptions** can blast pyroclasts weighing one ton or more up into the air at the speed of a jet plane.

- **Cinders and lapilli** are types of small pyroclasts. Cinders are 2.5–12 inches in diameter, lapilli are about 2.5 in.

- **Volcanic bombs** are blobs of molten magma that cool and harden as they fly through the air.

- **"Breadcrust" bombs** have a hard crust. The expanding hot center makes the crust crack, like a loaf of bread.

- **Around 90 percent** of the material ejected by explosive volcanoes is not lava, but tephra such as ash.

DID YOU KNOW?

Pumice rock is made from hardened lava froth. It is so full of air bubbles that it floats.

Hot-spot volcanoes

- **About 5 percent** of volcanoes are not near the margins of tectonic plates. Instead, they are thought to be over especially hot regions in the Earth's interior called hot spots.

- **Hot spots** are created by mantle plumes—hot currents that rise all the way from the core, through the mantle.

- **Mantle plumes** are about 300–600 mi across. They rise from as deep as 1,800 mi and melt their way through the crust to create hot-spot volcanoes.

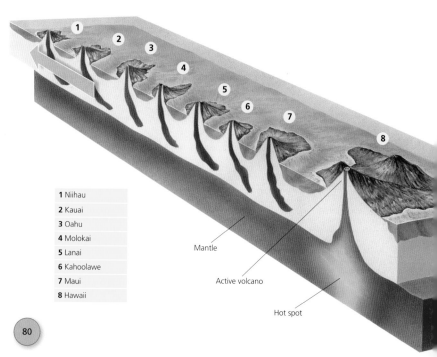

1 Niihau	
2 Kauai	
3 Oahu	
4 Molokai	
5 Lanai	Mantle
6 Kahoolawe	
7 Maui	Active volcano
8 Hawaii	

Hot spot

- **Famous hot-spot volcanoes** include the Hawaiian Islands in the Pacific Ocean, and the St. Helena chain of seamounts (mountains that are completely submerged underwater) in the Atlantic Ocean.

- **Hot-spot volcanoes** ooze runny lava that spreads out to create shield volcanoes.

- **Hot spots stay** in the same place while the plates slide over the top. At intervals, as the plate moves, a new volcano is created.

- **The Hawaiian Islands** are at the end of a chain of old volcanoes 3,700 mi long. The chain starts with the Meiji Seamount, which lies northeast of Japan.

- **The geysers**, hot springs, and bubbling mud pots of Yellowstone National Park, U.S.A, indicate a hot spot below.

- **Yellowstone has erupted** three times in the past two million years. The first produced over 2,000 times as much lava as the eruption of Mount St. Helens in 1980.

- **An alternative idea** is that hot spots are not unusually hot. They are places where the lithosphere is stretched thinner due to tectonic movements, and ordinary mantle material rises through.

◄ Where the Pacific plate has gradually moved over a hot spot, the Hawaiian Island chain has formed.

Tectonic plate

Earthquakes

- **Earthquakes are a shaking of the ground**. Some are slight tremors that barely rock a cradle. Others are so violent they can raze cities and mountains to the ground.

- **Small earthquakes** may be set off by landslides or volcanic eruptions. Large earthquakes are triggered by the grinding together of the tectonic plates that make up the Earth's surface.

▼ *During an earthquake, shock waves radiate in circles outward and upward from the focus of the earthquake. The damage caused is greatest at the epicenter, where the waves are strongest, but vibrations may be felt more than 600 mi away.*

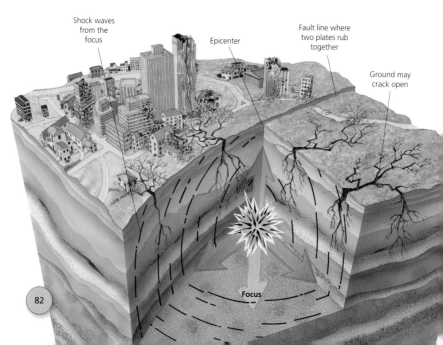

Shock waves from the focus

Epicenter

Fault line where two plates rub together

Ground may crack open

Focus

- **Tectonic plates** are sliding past each other all the time, but sometimes they get stuck. This causes rock to bend and stretch, until it snaps. This makes the plates jolt, sending out shock waves that cause quakes that can be felt far away.

- **Typically, tectonic plates slide** 1 or 2 inches past each other in a year. In a slip that triggers a major quake they can slip more than 3 ft in a few seconds.

- **In most earthquakes** a few minor tremors (foreshocks) are followed by an intense burst lasting just one or two minutes. A second series of minor tremors (aftershocks) occurs over the following few hours.

- **The starting point** of an earthquake below ground where most energy is released is called the focus, or hypocenter.

- **The epicenter** of an earthquake is the point on the surface directly above the focus.

- **An earthquake's effects** are usually strongest at the epicenter and become gradually weaker farther away.

- **Regions called earthquake zones** are especially prone to earthquakes. Most of these zones lie along the edges of tectonic plates.

- **A shallow earthquake** originates 0–43 mi below the ground, and these do the most damage.

- **An intermediate quake** begins 43–180 mi down.

- **Deep quakes** begin over 180 mi down. The deepest ever recorded earthquakes originated more than 370 mi below the surface.

Earthquake damage

▲ This bridge in Santiago, Chile, collapsed after an earthquake with a magnitude of 8.8 hit on February 27, 2010. At least 82 people were killed.

- **Some of the world's major cities** are located in earthquake zones, such as Los Angeles in the U.S.A, Mexico City in Mexico, and Tokyo in Japan.

- **Severe earthquakes** can cause buildings to collapse and rip up roads, rail lines, bridges, and tunnels.

- **When freeways collapsed** in the 1989 earthquake in San Francisco, U.S.A, some cars were crushed to just 20 in thick.

- **The 1906 earthquake** in San Francisco destroyed 250 mi of railway track around the city.

- **Some of the worst earthquake damage** is caused by fire, often set off by damage to gas and oil pipes and electrical cables.

- **In 1923**, 200,000 people died in the firestorm that engulfed the city of Tokyo, Japan. The quake hit at noon, toppling thousands of domestic fires lit ready to cook lunch.

- **In the Kobe earthquake** of 1995 in Japan, and the San Francisco earthquake of 1989, some of the worst damage was to buildings built on landfill—loose material thrown into the ground to build up the land.

- **The earthquake that killed** the most people is believed to have hit Shaanxi, China, in 1556. It is estimated to have claimed 830,000 lives.

DID YOU KNOW?
There are probably over one million earthquakes each year, but fewer than 25,000 are detected by scientific instruments.

Earthquake measurement

- **Earthquakes are measured** with a device called a seismometer (seismograph).

- **The Richter scale** measures the magnitude (size) of an earthquake on a scale of one to ten. Each step up indicates a tenfold increase in energy.

- **The Richter scale** was devised in the 1930s by American geophysicist Charles Richter (1900–1985).

- **The Modified Mercalli scale** assesses a quake's severity according to its effects.

- **The Mercalli scale** was devised by Italian scientist Giuseppe Mercalli (1850–1914).

- **A Mercalli scale I earthquake** is almost undetectable. A Mercalli scale XII earthquake causes almost total destruction.

- **The modern measuring scale** called the MMS—moment magnitude scale—combines Richter readings with observations of rock movements and measures the amount of energy released. Its numbers are similar to the Richter scale.

- **The most powerful earthquake** ever recorded was the Valdivia earthquake in Chile on May 12, 1960, measuring 9.5 on the moment magnitude scale.

- **The Indian Ocean earthquake** of December 2004, which set off a tsunami, measured Richter 9.3.

- **Between 10–20 earthquakes** a year reach 7 on the Richter scale.

MERCALLI SCALE	
I. Instrumental	Only detected by seismographs.
II. Weak	Felt by only a few people, especially on upper floors.
III. Slight	Felt by some people, especially indoors and on upper floors.
IV. Moderate	Felt by people indoors, some outdoors. Windows and doors rattle. Standing cars rock.
V. Rather strong	Felt by most people. People sleeping wake, small unstable objects fall over. Trees shake.
VI. Strong	Felt by everyone. Difficult to walk. Heavy objects moved. Structural damage is slight.
VII. Very strong	Difficult to stand up. Slight to moderate damage in well-built urban areas. Moderate damage to poorly built structures.
VIII. Destructive	Considerable damage to ordinary buildings, some may partially collapse. Severe damage to poorly built structures.
IX. Violent	Damage great in ordinary buildings, with partial collapse. Buildings shifted off foundations.
X. Intense	Some well-built wooden structures destroyed. Most stone and frame structures destroyed with foundations. Train rails bent.
XI. Extreme	Few structures remain standing. Bridges destroyed. Train rails bent greatly.
XII. Cataclysmic	Damage total. Ground moves in waves and the landscape is altered. Objects thrown into the air.

▲ The Mercalli scale evaluates the severity of a quake from the damage it does on a scale of one–12 (I–XII).

Earthquake waves

- **Earthquake waves** are the vibrations sent out through the ground by earthquakes. They are also called seismic waves.

- **There are two kinds** of deep earthquake wave: primary (P) waves and secondary (S) waves.

- **P waves** travel at 3.1 mi per second and move by alternately squeezing and stretching rock.

- **S waves** travel at 1.8 mi per second and move the ground up and down or from side to side.

- **There are two kinds** of surface earthquake wave: Love (Q) waves and Rayleigh (R) waves.

▼ *This shows how the ground is vibrated by waves underground (P and S waves) and on the surface (Q and R waves).*

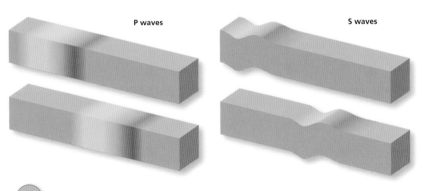

P waves

S waves

- **Love waves** shake the ground from side to side in a jerky movement that may topple tall buildings.

- **Rayleigh waves** shake the ground up and down, making it seem to roll.

- **In solid ground**, earthquake waves travel too fast to be seen. However, they can turn loose sediments into a fluidlike material. Then earthquake waves can be seen rippling across the ground like waves in the sea.

DID YOU KNOW?
Some earthquake waves travel at 20 times the speed of sound.

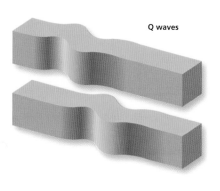

Q waves

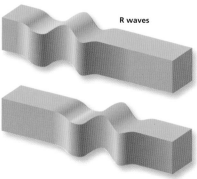

R waves

Earthquake prediction

- **One way to predict** earthquakes is to study past quakes.

- **If there has been no earthquake** in a prone area for a while, it is more likely there will be one soon. The longer it has been, the bigger the next earthquake will be.

- **Seismic gaps** are places in active earthquake zones where there has been no earthquake activity. This is where a severe earthquake is more likely to occur.

- **Seismologists use ground instruments** and laser beams bounced off satellites to detect tiny distortions in the rock, which may indicate a build up of strain.

- **A linked network** of four laser-satellite stations called Keystone monitors ground movements in Tokyo Bay, Japan, so that earthquakes can be predicted.

- **The level of water** in the ground may indicate strain. The rock can squeeze groundwater toward the surface.

- **Rising surface levels** of the underground gas radon may also indicate that the rock is being squeezed.

- **Other signs of strain** in the rock may show as changes in the ground's electrical resistance or its magnetism.

- **Before an earthquake**, dogs are said to howl, rats and mice scamper from their holes, and fish thrash about.

▶ *This image of the Hayward Fault, California, uses satellite pictures and radar measurements. It shows the ground movement between 1992 and 1997 (orange and red indicate the most movement).*

Famous earthquakes

- **The palaces of the Minoan people** on the island of Crete were destroyed by an earthquake and tsunami around 1650 BC.

- **The earliest well-documented earthquake** hit the ancient Greek town of Sparta in 464 BC, killing 20,000 people.

▼ *In January 2010, a powerful earthquake hit the city of Port-au-Prince in Haiti, in the Caribbean, with devastating effects. Thousands of buildings collapsed.*

- **In July 1201** an earthquake rocked every city in the eastern Mediterranean. It is estimated to have killed up to one million people.

- **In 1556** an earthquake thought to have been 8.3 on the Richter scale hit the province of Shaanxi in China.

- **In 1906,** the U.S. city of San Francisco was shaken by an earthquake that lasted three minutes. The earthquake started fires that burned the city almost to the ground.

- **The 1923 earthquake** that devastated the Japanese cities of Tokyo and Yokohama caused the seabed in nearby Sagami Bay to drop by over 1,300 ft.

- **The 1755 Lisbon earthquake** prompted the French writer Voltaire to write *Candide*, a book that inspired the French and American revolutions.

- **The Tangshan earthquake** of 1976 in China was the deadliest of the 20th century. It killed over 250,000 people, as the entire city of Tangshan was leveled.

- **The powerful 7.0 magnitude** earthquake that hit Haiti on January 12, 2010 took the lives of up to 230,000 people and made more than one million people homeless.

- **In 2013** a 7.7 magnitude quake hit Balochistan, Pakistan and killed over 800 people.

Tsunamis

- **Tsunamis are huge waves** that are triggered when the seafloor is shaken violently by an earthquake, landslide, or volcanic eruption.

- **In deep water**, tsunamis travel almost unnoticed below the surface. However, once they reach shallow coastal waters they rear up into waves 98 ft or higher.

- **Often called "tidal waves,"** tsunamis are nothing to do with tides. The word tsunami (pronounced "soo-nah-mee") is Japanese for "harbor wave."

- **Tsunamis usually come** in a series of a dozen or more waves, anything from five minutes to one hour apart.

- **Before a tsunami hits**, the sea may recede, moving back away from the beach.

- **Tsunamis can travel** along the seabed as fast as a jet plane, at 435 mph or more.

- **On December 24, 2004**, an earthquake beneath the sea off Sumatra, Indonesia, generated a huge tsunami that spread across the Indian Ocean. It is now known as the Asian or Indian Ocean tsunami.

- **The Asian tsunami** was 98 ft high and traveled 5,000 mi to South Africa. Over 225,000 people were killed.

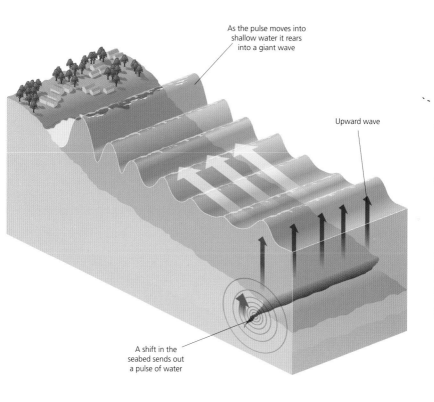

As the pulse moves into shallow water it rears into a giant wave

Upward wave

A shift in the seabed sends out a pulse of water

▲ *Tsunamis may be generated underwater by an earthquake, then travel along the seabed before emerging to inundate a coast.*

Shaping the land

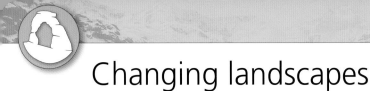

Changing landscapes

- **The Earth's surface** changes all the time. Most changes take millions of years, but sometimes the landscape is reshaped suddenly by an avalanche, earthquake, or volcano.

- **The huge forces** of the Earth's interior distort and reform the Earth's surface from below. Weather, water, ice, and other "agents of erosion" mold it from above.

- **Most landscapes**, except deserts, are molded by running water, which explains why hills have rounded slopes. Dry landscapes are more angular, but even in deserts water often plays a major shaping role.

- **Mountain peaks** are jagged because the extreme cold high up causes the rocks to freeze and shatter.

- **American scientist W. M. Davis** (1850–1935) thought that landscapes are shaped by repeated "cycles of erosion."

- **Davis's cycles of erosion** have three stages: Vigorous "youth," steady "maturity," and sluggish "old age."

- **Observation has shown** that erosion does not become slower as time goes on, as Davis believed.

- **Many landscapes** that exist today have been shaped by forces no longer in operation, or that are now much smaller than they were, such as the moving ice of glaciers during past ice ages.

◄ In southwestern U.S.A., the Colorado Plateau has uplifted and the Colorado River has cut down into it to form a deep canyon.

Weathering

- **Weathering is the gradual breakdown** of rocks when they are exposed to air, water, and living things.

- **Weathering affects** surface rocks the most, but water trickling into the ground can weather rocks 650 ft down.

- **The more extreme the climate**—very cold or very hot—the faster weathering takes place.

- **In tropical Africa**, the basal weathering front (where weathered meets unweathered rock) is often more than 190 ft down.

- **Weathering works chemically** (through chemicals in rainwater), mechanically (through temperature changes), and organically (through plants and animals).

- **Chemical weathering** is when gases such as carbon dioxide and sulfur oxides dissolve in rain to form weak acids that corrode rocks such as limestone.

- **The main form** of mechanical weathering is frost shattering. This is when water expands as it freezes in cracks and so splits apart and shatters the rock.

Thermoclastis is when desert rocks crack as they get hot and
expand in the day, then cool and contract at night.

▶ The combination of
extreme temperatures
and the damaging
effect of strong winds
has created this rock
sculpture in the
Arizona desert, U.S.A.

DID YOU KNOW?
At -76°F, ice can exert a
pressure of three tons on an
area of rock the size of a
postage stamp.

101

Limestone weathering

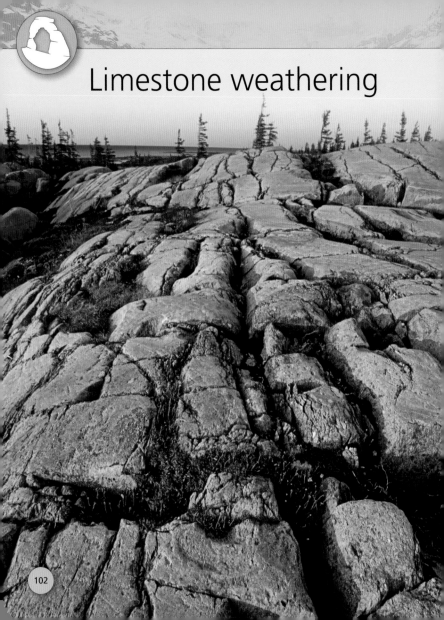

- **Streams and rainwater** absorb carbon dioxide gas from soil and air, turning them into weak carbonic acid.

- **Carbonic acid corrodes** (wears away by dissolving) limestone in a process called carbonation.

- **When limestone rock** is close to the surface, carbonation can create spectacular scenery.

- **Corroded limestone** scenery is often called karst. This is because a very good example of it is the Karst Plateau near Dalmatia, in Bosnia.

- **On the surface**, carbonation eats along cracks to create "pavements," with slabs called clints. The slabs are separated by deeply etched grooves called grykes.

- **Limestone rock** does not soak up water like a sponge. It has massive cracks called joints, and streams and rainwater trickle deep into the rock through these cracks.

- **Streams drop down** into limestone through sinkholes, like water down a plughole. Carbonation eats out such holes to form giant shafts called potholes.

- **Some potholes** are eaten out to create great funnel-shaped hollows called dolines, up to 330 ft across.

- **Where water streams** out along horizontal cracks at the base of potholes, the rock may be etched out into caverns.

- **Caverns may be eaten out** so much that the roof collapses to form a gorge or a large hole called a polje.

◄ *The acidity of rainwater has etched out the cracks in limestone to create a limestone "pavement" on The Burren, Ireland.*

Caves

- **Caves are giant holes** that run horizontally underground. Holes that plunge vertically are called potholes.

- **The most spectacular caves,** called caverns, are found in limestone. Acid rainwater trickles through cracks in the rock and wears away huge cavities.

- **The world's largest known single cave** is the Sarawak Chamber in Gunung Mulu in Sarawak, Malaysia.

- **The deepest cave** yet found is Krubera Cave in Georgia, 7,200 ft below the surface.

- **The longest cave system** is the Mammoth Cave in Kentucky, U.S.A, which is 400 mi long and not yet fully explored.

- **Many caverns** contain speleothems. These deposits are made mainly from calcium carbonate deposited by water trickling through the cave.

- **Stalactites** are icicle-like speleothems that hang from cave ceilings. Stalagmites grow upward from the floor.

- **The world's longest stalactite** is 27 ft in Jeita Grotto, Lebanon.

- **The world's tallest stalagmite** at 229 ft is in Son Doong Cave, Vietnam.

> **DID YOU KNOW?**
> The Sarawak Chamber is big enough to hold the world's biggest stadium—the 150,000-capacity Rungrado 1 May Day Stadium, North Korea—with room to spare.

▶ *Carlsbad caverns, New Mexico, U.S.A. Caverns can be subterranean palaces filled with glistening pillars.*

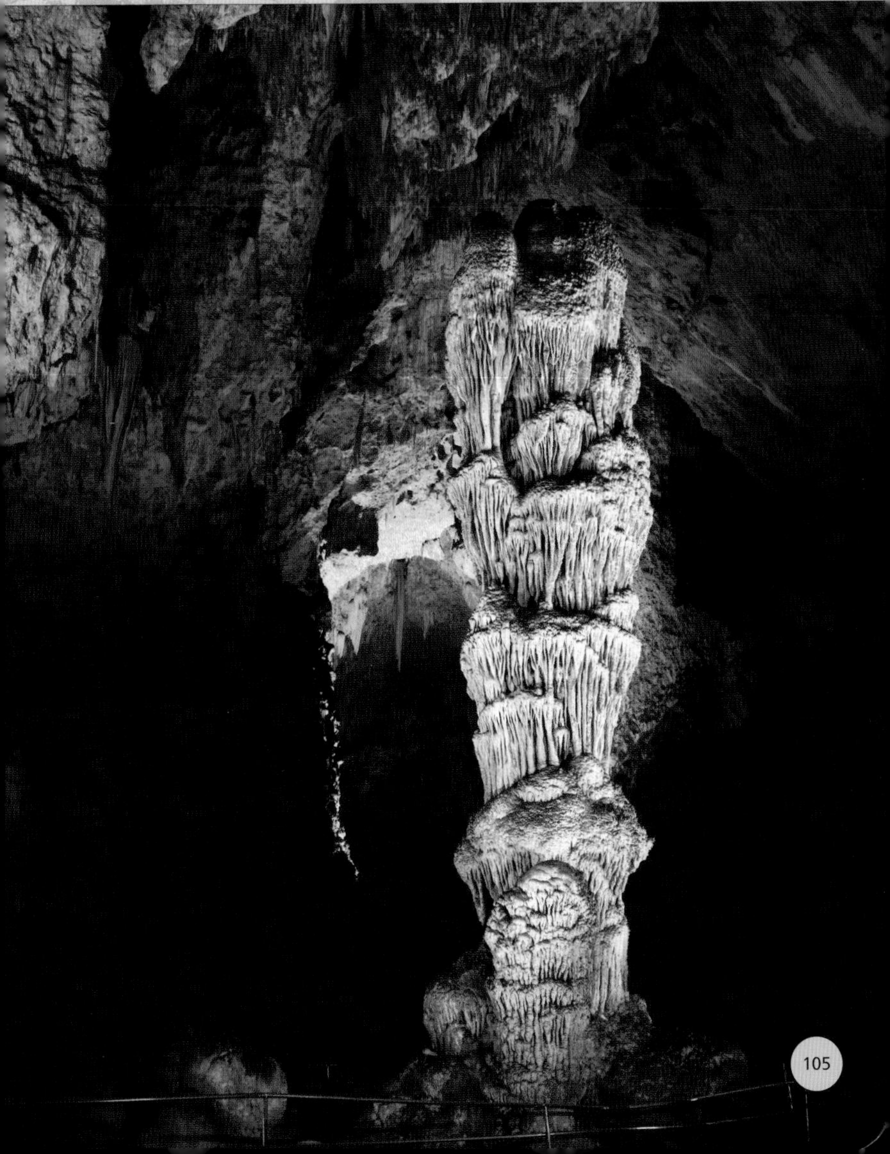

Rivers

● **The water that fills rivers** comes from rainfall running directly off the land, from melting snow or ice, or from a spring bubbling out water that has soaked into the ground.

● **High up in mountains** near their sources (starts), rivers are usually small and steep. They tumble over rocks through narrow valleys, which they have carved out over thousands of years.

● **All the rivers** in a particular area, called a catchment area, flow down to join each other, like branches on a tree. The branches are called tributaries. The bigger the river, the more tributaries it is likely to have.

● **As rivers flow downhill**, they are joined by more tributaries and grow bigger.

● **In its lower reaches** a river is often wide and deep. It winds back and forth in meanders (bends) across floodplains made of silt.

● **Rivers flow fast** over rapids in their upper reaches.

● **The banks and beds** of rivers are worn away by gravel and sand, and the force of the moving water.

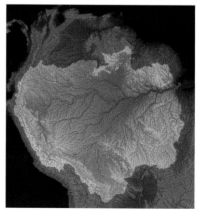

▲ *A computerized satellite image reveals the fantastic dendritic, or branching, network of the river Amazon and its tributaries.*

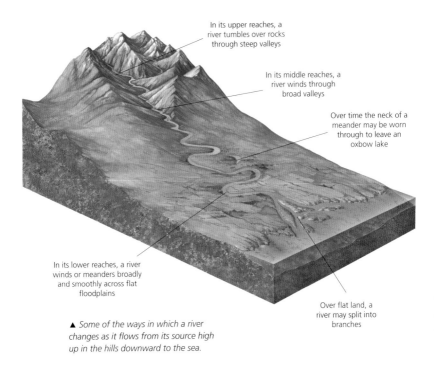

In its upper reaches, a river tumbles over rocks through steep valleys

In its middle reaches, a river winds through broad valleys

Over time the neck of a meander may be worn through to leave an oxbow lake

In its lower reaches, a river winds or meanders broadly and smoothly across flat floodplains

Over flat land, a river may split into branches

▲ Some of the ways in which a river changes as it flows from its source high up in the hills downward to the sea.

● **Every river carries sediment**. This consists of sand, large stones that are rolled along, and fine silt that is suspended in the water.

● **The discharge** of a river is the amount of water flowing past a certain point each second (in cubic feet per second).

● **Intermittent rivers** only flow after rain. Rivers that flow all year round are perennial—they are fed by water flowing underground, as well as rain.

Hills

- **Mountains are solid rock**, but hills can be solid rock or piles of debris built up by glaciers, sand, or the wind.

- **Hills made of solid rock** are either very old, having been worn down from mountains over millions of years, or they are made from soft sediments that were low hills.

- **In moist climates**, hills are often rounded by weathering and by water running over the land.

▼ The contours of hills in damp places have often been gently rounded over long periods by a combination of weathering and erosion by running water.

- **As solid rock is weathered**, the hill is covered in a layer of debris called regolith, including broken rock, soil, dust, and other material.

- **Hills often have** a shallow S-shaped slope, called a "convexo-concave" slope. There is a rounded convex shape at the top, and a long concave slope lower down.

- **Hill slopes become gentler** as they are worn away because the top is worn away faster. This is called decline.

- **Some hill slopes stay equally steep**, but are simply worn back. This is called retreat.

- **Some hill slopes wear backward** as gentler sections get longer and steeper sections get shorter. This is called replacement.

River valleys

▼ *Rivers carve out valleys over hundreds of thousands of years as they grind material along their beds.*

- **Rivers carve out valleys** as they wear away their channels.

- **High up in the mountains**, much of a river's energy is spent on carving into or eroding the riverbed. The valleys there are deep, with steep sides.

- **Farther down toward the sea**, more of a river's erosive energy goes into wearing away its banks. It carves out a broader valley as it winds back and forth.

- **Large meanders** normally develop only when a river is crossing broad, flat plains in its lower reaches.

- **Incised meanders** are those that have been carved into deep valleys. These meanders form when a river flows across a low plain. The plain is lifted up and the river cuts down into it, keeping its meanders.

- **The Grand Canyon** is made of incised meanders. They were created as the Colorado River cut into the Colorado Plateau after it was uplifted 17 million years ago.

- **Some valleys** seem far too big for their river alone to have carved them. Such a river is "underfit" or "misfit."

- **Many large valleys** with misfit rivers were carved out by glaciers or glacial meltwaters.

- **The world's rivers** wear the entire land surface down by an average of 3 in every 1,000 years.

River channels

- **A channel** is the long trough along which a river flows, often at the base of a river valley.

- **When the channel of a river** winds or has a rough bed, friction slows the river down.

▼ *Where the land is flatter, rivers wind and often divide into separate channels.*

- **A river flows faster** through a narrow, deep channel than a wide, shallow one because there is less friction.

- **All river channels tend to wind**, and the nearer they are to sea level, the more they wind. They form remarkably regular meanders.

DID YOU KNOW?
Meanders can form almost complete loops with only a neck of land separating the ends.

- **Meanders seem to develop** because of the way in which a river erodes and deposits sediments.

- **One key factor in meanders** is the ups and downs along the river, called pools (deeps) and riffles (shallows).

- **The distance between pools and riffles**, and the size of meanders, are in close proportion to the river's width.

- **Another key factor in meanders** is the tendency of river water to flow not only straight downstream but also across the channel. Water spirals through the channel in a corkscrew fashion called helicoidal flow.

- **Helicoidal flow** makes water flow faster on the outside of bends, wearing away the banks. It flows more slowly on the inside, building up deposits called slip-off slopes such as sandbars or mudflats.

Waterfalls

- **When a river plunges vertically**, it is called a waterfall.

- **Waterfalls may form** where the river flows over a band of hard rock, such as a volcanic sill. The river erodes the soft rock below but has little effect on the hard band.

- **If a stream's course** is suddenly broken, for example if it flows over a cliff into the sea, over a fault or over a hanging valley, a waterfall can form.

- **Boulders often swirl** around at the foot of a waterfall, wearing out a deep pool called a plunge pool.

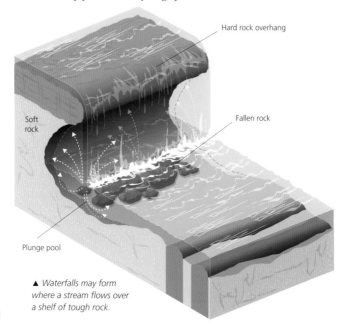

Hard rock overhang

Fallen rock

Soft rock

Plunge pool

▲ *Waterfalls may form where a stream flows over a shelf of tough rock.*

- **Angel Falls** in Venezuela are named after American pilot Jimmy Angel who flew over them in 1935. They have the longest straight drop or plunge at 2,648 ft.

- **Victoria Falls** in Zimbabwe, Africa, are 360 ft high and known locally as *Mosi oa Tunya*, which means the "smoke that thunders."

- **The roar from Victoria Falls** can be heard 25 mi away.

- **Niagara Falls** on the U.S./Canadian border developed where the Niagara River flows out of Lake Erie.

- **Niagara consists of two falls**: Horseshoe Falls, 177 ft high, and American Falls, 180 ft high.

▼ *In the wet season, over 106,000 cu ft of water flow over Zambia/Zimbabwe's Victoria Falls every second.*

Floods

- **A flood** occurs when a river or the sea rises so much that it spills over the surrounding land.

- **River floods** may occur after a period of prolonged heavy rain or after snow melt in spring.

- **Small floods** are common, but big floods are rare. So flood size is described in terms of frequency.

- **A two-year flood** is a smallish flood that is likely to occur every two years. A 100-year flood is a big flood that is likely to occur once each century.

- **A flash flood** occurs when a small stream or even dry stream bed changes to a raging torrent after heavy rain during a dry spell.

- **The 1993 flood** on the Mississippi–Missouri caused damage of $15 billion and made 75,000 people homeless, despite massive flood control works in the 1930s.

- **The Hwang Ho** or Yellow River in China is also known as "China's sorrow" because its floods are so devastating. In 1887 about one million people died in a flood there.

- **In July/August 2010**, widespread flooding, started by heavy monsoon rains in northern Pakistan, made many millions of people homeless.

- **In 2013** an estimated 6,000 people died in floods in Northern India.

- **Not all floods are bad**. Before the Aswan Dam was built, Egyptian farmers relied on the yearly flooding of the Nile to enrich the soil.

▲ Even when people are rescued, a flood
can destroy homes and wash away soil
from farmland, leaving it barren.

Swamps and marshes

▼ Mangrove swamps are created by the unique
ability of mangrove trees to live in salt water

- **Wetlands are areas of land** where the water level is mostly above the ground.

- **The main types of wetland** are bogs, fens, swamps, and marshes.

- **Bogs and fens** occur in cold climates and contain plenty of partially rotted plant material called peat.

- **Marshes and swamps** are found in warm and cold places. They have more plants than bogs and fens.

- **Marshes are in** permanently wet places, such as shallow lakes and river deltas. Reeds and rushes grow in marshes.

- **Swamps develop** where the water level varies—often along the edges of tropical rivers where they are flooded, notably along the Amazon and Congo. Trees such as mangroves grow in swamps.

- **Half the wetlands in the U.S.A.** were drained before their value was appreciated. Almost half of the 400 sq mi area of Dismal Swamp, North Carolina, has been drained.

- **The Pripet Marshes** on the borders of Belorussia are the biggest in Europe, covering 104,000 sq mi.

- **Floods are controled by wetlands** since they act like sponges, soaking up heavy rain then releasing the water slowly.

- **Wetlands also act to top up supplies** of groundwater and even have an effect on the local climate, helping to reduce the extremes of heat and cold.

Lakes

- **Most of the world's large lakes** lie in regions that were once glaciated. The glaciers carved out deep hollows in the rock in which water collected. The large lakes of the U.S.A. and Canada are partly glacial in origin.

- **In Minnesota**, U.S.A., 11,000 lakes were formed by glaciers.

- **The world's deepest lakes** are often formed by faults in the Earth's crust, such as Lake Baikal in Siberia and Lake Tanganyika in East Africa.

- **Most lakes last** only a few thousand years before they are filled in by silt or drained by changes in the landscape.

- **The world's largest lake** is the Caspian Sea, which is a saltwater lake.

- **The world's highest large lake** is Titicaca in South America, which is 12,500 ft above sea level.

- **The world's lowest large lake** is the Dead Sea between Israel and Jordan. It is 1,300 ft below sea level and getting lower all the time.

- **The largest underground lake** in the world is Drauchen-hauchloch or Dragon's Breath, which is inside a cave in Namibia.

- **The Aral Sea** in West Asia shrank to only one-tenth of its original size, which was 27,000 sq mi, between 1950 and 2000, due to diverting the rivers that fed it to water crops elsewhere.

- **New dam and pipeline projects** hope to refill part of the Aral Sea over the next 50 years.

▶ *About 50 years ago the Aral Sea would have almost filled this photograph's frame. By 2014 it was reduced to two main and several smaller lakes. The whole south and east have become desert.*

Deserts

- **Deserts are dry places** where it rarely rains. Many are hot, but one of the biggest deserts is Antarctica. Deserts cover about one-fifth of the Earth's land.

- **Desert that is strewn** with boulders is called hamada. Desert that is blanketed with gravel is called reg.

- **About one-fifth** of all deserts consist of sand dunes. These are known as ergs in the Sahara.

- **The type of sand dune** depends on how much sand there is, and how changeable the wind is.

- **Barchans are moving**, crescent-shaped dunes that form in sparse sand where the wind direction is constant.

- **Seifs are long dunes** that form where sand is sparse and the wind comes from two or more directions.

▶ Sand dunes are sculpted and piled up into different shapes by the wind.

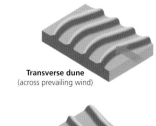

Transverse dune
(across prevailing wind)

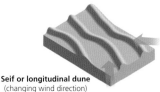

Seif or longitudinal dune
(changing wind direction)

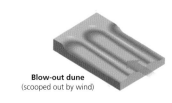

Barchan dune
(relatively constant wind direction)

Blow-out dune
(scooped out by wind)

Star dune
(variable wind direction)

- **Most streams in deserts** flow only occasionally, leaving dry stream beds called wadis or arroyos. These may fill suddenly to become a flash flood site after rain.

- **In cool, wet regions**, hills are covered in soil and rounded in shape. In deserts, hills are bare rock with cliff faces supported by straight slopes.

- **Mesas and buttes** are pillarlike plateaus that have been carved gradually by water in deserts.

▼ *In the south of Morocco, in the Western Sahara, there are great seas of sand called ergs.*

Glaciers

- **Glaciers are "rivers"** of slowly moving ice. They form in mountain regions when it is too cold for snow to melt. They flow down through valleys, creeping lower until they melt in the warm air lower down.

- **When new snow**, or névé, falls on old snow, glaciers are formed. The weight of the new snow compacts the old into denser snow called firn.

- **In firn snow**, all the air is squeezed out so it looks like white ice. As more snow falls, firn gets more compacted and turns into glacier ice that slides slowly downhill.

- **Today, glaciers form** only in high mountains and toward the North and South Poles. In the ice ages, glaciers were widespread and left glaciated landscapes in many places that are now free of ice.

- **As glaciers move downhill**, over humps and around bends, they bend and stretch, opening up deep cracks called crevasses. Sometimes these occur where the glacier passes over a ridge of rock.

- **The biggest crevasse** is often called the bergschrund. It forms when the ice pulls away from the back wall of the hollow where the glacier starts.

- **The underside of a glacier** is warmish (about 32°F). It moves by gliding over a film of water that forms as pressure melts the glacier's base. This is called basal slip.

- **Where the underside** of a glacier is well below 3°F, it moves as if layers were slipping over each other like a pack of cards. This is called internal deformation.

- **Valley glaciers** are glaciers that flow in existing valleys.

- **Cirque glaciers** are small glaciers that flow from hollows high up.

- **Alpine valley glaciers** form when several cirque glaciers merge.

- **Piedmont glaciers** form where valley glaciers join as they emerge from the mountains.

▼ *Glaciers begin in small hollows in the mountain called cirques, or corries. They flow downhill, gathering huge piles of debris called moraine on the way.*

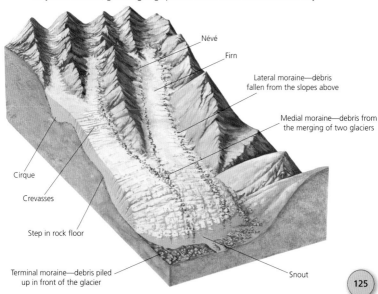

Névé

Firn

Lateral moraine—debris fallen from the slopes above

Medial moraine—debris from the merging of two glaciers

Cirque

Crevasses

Step in rock floor

Terminal moraine—debris piled up in front of the glacier

Snout

Glaciated landscapes

- **Glaciers move slowly** but their sheer weight and size give them enormous power to shape the landscape.

- **Over tens of thousands of years** glaciers carve out winding valleys into huge, straight U-shaped troughs (river valleys are usually V-shaped).

▼ After an ice age, glaciers leave behind a dramatically altered landscape of deep valleys and piles of debris.

- **Glaciers may truncate** (slice off) tributary valleys to leave them "hanging," with a cliff edge high above the main valley. Hill spurs (ends of hills) may also be truncated.

DID YOU KNOW?

Some glaciers, such as Jakobshavn Glacier in Greenland, have moved faster than 98 ft per day.

- **Cirques, or corries**, are armchair-shaped hollows carved out where a glacier begins high up in the mountains.

- **Arêtes are knife-edge ridges** that are left between several cirques as the glaciers in them cut backward.

- **Drift is a blanket of debris** deposited by glaciers. Glaciofluvial drift is left by the water made as the ice melts. Till is left by the ice itself.

- **Drumlins** are egg-shaped mounds of till.

- **Eskers** are snaking ridges of drift left by streams under the ice.

- **Moraine** is piles of debris left by glaciers as they melt and retreat.

- **One of the main effects** of the global warming happening today is to make glaciers shrink, become shorter, or even disappear.

- **Some major glaciers** are now 1–3 mi shorter than they were 50 years ago.

Cold landscapes

- **"Periglacial" is used to describe** conditions next to the ice in the ice ages. It now means similar conditions found today.

- **Periglacial conditions** are found on the tundra of northern Canada and Siberia and on nunataks, which are the hills that protrude above ice sheets and glaciers.

- **In periglacial areas**, ice melts only in spring at the surface.

- **Where soil stays frozen** for almost the whole year, usually just under the surface, it is known as permafrost.

- **When the ground above** the permafrost melts, the soil twists into buckled layers called involutions.

- **When frozen soil melts** it becomes so fluid that it can creep easily down slopes, creating large tongues and terraces.

- **Frost heave** is the process when frost pushes stones to the surface as the ground freezes.

- **After frost heave**, large stones roll downhill and away, leaving the fine stones on top. This creates intricate patterns on the ground.

- **On flat ground**, quiltlike patterns are called stone polygons. On slopes, they stretch into stone stripes.

- **Pingos are mounds** of soil with a core of ice. They are created when groundwater freezes beneath a lake.

▼ Cold conditions such as on this Arctic tundra create a unique landscape in polar and high mountain regions.

DID YOU KNOW?
In periglacial conditions, temperatures never climb above freezing in winter.

Continents and oceans

Earth's continents

- **The word "continent"** means continuous or connected, and a continent is loosely defined as a continuous landmass surrounded by a stretch of water.

- **A continent** can also be defined geologically by rock types and structures, such as America or Afro-Eurasia. It can also be defined geographically, such as South America or Asia.

- **Some definitions recognize** four continents: Afro-Eurasia, America, Antarctica, and Australia.

- **Others recognize five continents**: Africa, Eurasia, America, Antarctica, and Australia.

- **Some definitions have six continents**: Africa, Eurasia, North America, South America, Antarctica, and Australia.

- **Many common descriptions** have seven different continents: Africa, Europe, Asia, North America, South America, Antarctica, and Australia.

- **Australia is sometimes included** with New Zealand and the Southwest Pacific Islands as a continent called Oceania.

- **Supercontinents such as Pangaea**, then Laurasia and Gondwana, existed in the past when the present continents were joined together.

- **Subcontinents refer to areas** of continents which, although joined, are on separate tectonic plates, for example the Indian subcontinent to the south of Asia.

● **The geological definition** of a continent refers to the Earth's crust, with thick continental crust built around a core, or craton, of very ancient, stable rocks.

▼ *Some 220 million years ago all the present continents were joined and surrounded by a massive ocean, Panthalassa. By 150 million years this divided into northern and southern supercontinents with the Tethys Sea opening between them.*

Pangaea (left) was composed of all the modern continents (right).

As Laurasia drifted from Gondwana (left), the Tethys Sea widened (right).

133

Europe

- **The smallest continent**, Europe has an area of 4,000,000 sq mi. For its size, Europe has a very long coastline, over 43,000 mi.

- **In the north** are the ancient glaciated mountains of Scandinavia and Scotland, which were once much higher.

- **Across the center** are the lowlands of the North European Plain, stretching from the Urals in Russia to France in the west.

- **Much of southern Europe** has been piled up into young mountain ranges like the Alps, as Africa drifts slowly north.

- **The highest point** in Europe is Mount Elbrus in the Russian Caucasus, at 18,510 ft high.

- **Northwest Europe** was once joined to Canada. The ancient Caledonian mountains of eastern Canada, Greenland, Scandinavia, and Scotland were formed together as a single mountain chain 360–540 million years ago.

- **Mediterranean Europe** has warm summers and mild winters.

- **Northwest Europe** is often wet and windy. It has mild winters because it is bathed by the warm North Atlantic Drift.

- **The Russian islands** of Novaya Zimlya are far into the Arctic Circle and are icebound in winter.

- **The largest lake** is Ladoga in Russia, at 6,795 sq mi.

▼ *The Rock of Gibraltar in the far southwest of Europe marks the entrance to the Mediterranean Sea.*

Africa

- **Africa is the world's** second largest continent. It stretches from the Mediterranean in the north to the Cape of Good Hope in the south. Its area is over 11,600,000 sq mi.

- **It is the world's warmest continent**, lying almost entirely within the tropics or subtropics.

- **Temperatures in the Sahara** are among the highest on Earth, often soaring over 122°F.

- **The Sahara** in the north of Africa, and the Kalahari in the south, are the world's largest deserts. Most of the continent in between is savanna (grassland) and bush. In the west and center are lush tropical rain forests.

▼ Lying in the tropics, much of East Africa is grassland or savanna, too dry for half the year for trees to grow.

DID YOU KNOW?

Modern humans, Homo sapiens, first evolved in Africa about 200,000 years ago.

- **Much of Africa consists** of vast plains and plateaus, broken in places by mountains such as the Atlas range in the northwest and the Ruwenzori in the center.

- **The Great Rift Valley** runs 4,400 mi from the Red Sea. It is a huge trench in the Earth's surface opened up by the pulling apart of two giant tectonic plates.

- **At 26,600 sq mi**, Lake Victoria is Africa's largest lake.

- **Africa's highest mountain** is Kilimanjaro, at 19,341 ft.

- **The world's biggest sand dunes**, over 1,300 ft high, can be found in the Sahara at Erg Tifernine, Algeria.

Mount Elgon

Great Rift Valley

▲ *A satellite view of the Great Rift Valley in Kenya, with the cone of the extinct volcano Mount Elgon on the left.*

137

Asia

- **Asia is the world's largest continent**, stretching from Europe in the west to Japan in the east. It has an area of 17,220 sq mi.

- **Asia has huge climate extremes**, from a cold polar climate in the north to a hot tropical one in the south.

- **Verkhoyansk in Siberia** has had temperatures as high as 98.6°F and as low as −90.4°F

- **The Himalayas** are the highest mountains in the world, with 14 peaks over 26,000 ft high. To the north are vast, empty deserts, broad grasslands, and huge coniferous forests. To the south are fertile plains and valleys and steamy tropical jungles.

▼ With giant sand dunes in the distance, Bactrian (two-humped) camels graze on the dry grasses of Mongolia, between Russia and China at the heart of the Asian continent.

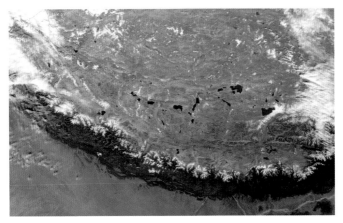

▲ *Sometimes called the "roof of the world," the Tibetan Plateau is on average over 14,700 ft high and covers an area four times the size of France. The Himalayas border the Tibetan Plateau on the south.*

- **Northern Asia** sits on one giant tectonic plate.

- **India is on a separate plate** that crashed into northern Asia around 50 million years ago. It is pushing up the Himalayas as it plows on northward.

- **Asia's longest river** is China's Yangtze, at over 3,700 mi long.

- **Asia's highest mountain** is also the highest in the world—Mount Everest, or Sagarmatha in Nepal, at 29, 028 ft

- **The Caspian Sea** between Azerbaijan and Kazakhstan is the world's largest lake, covering 146,100 sq mi.

DID YOU KNOW?

The place on Earth farthest from the ocean is in Xinjiang Province, northwest China, at more than 1,500 mi from an oceanic coastline.

Australia, island continent

- **The continent of Australia** includes all the lands that sit on the same continental shelf, such as Tasmania, New Guinea, and Seram (but not New Zealand).

- **Mainland Australia** (with the island of Tasmania) is the smallest of the world's seven continents, with a total land area of almost 3,000,000 sq mi.

- **Until about 95 million years ago** Australia was joined with Antarctica as the southern supercontinent called Gondwana.

- **Australia itself was connected** to Tasmania by dry land during the last Ice Age, about 20,000 years ago, when sea levels were much lower than today.

- **Australia only split** from New Guinea between 8,000 and 6,500 years ago when sea levels rose again.

- **The tectonic plate** carrying Australia is on the move and heading towards Eurasia at 2–2.7 in per year.

- **Australia lies right in the middle** of the Indo-Australian tectonic plate so it has "quiet" geology. For example, it is the only continent with no active or dormant volcanoes.

- **The continent has been isolated** from the rest of the world for 40 million years, during which time it drifted northward and became warmer and more arid.

- **Australia has woodland**, rain forest, and grassland areas but most of it is desert or semidesert. It is the driest inhabited continent.

- **Also the flattest continent**, the only mountains in Australia are the eastern Great Dividing Range, with few peaks over 6,500 ft.

- **Australia is a very ancient continent**, with all the main kinds of rocks, the oldest being 3.8 billion years old.

▼ *Uluru is a giant exposed sandstone rock formation in central Australia.*

Oceania

- **Oceania is a vast region** that includes Australia, New Zealand, New Guinea, and islands spread over much of the Pacific Ocean.

- **The land area is 3,290,000 sq mi**, with most of this being Australia, but the sea area is much bigger.

- **Apart from Australia**, the largest island is New Guinea, at 303,400 sq mi.

- **Oceania is mostly tropical**, with temperatures averaging 86°F in the north of Australia, and slightly lower on the islands where the ocean keeps the land cool.

▼ *In the center of New Zealand's North Island, Tongariro National Park has three active volcanoes, where the Indian-Australian tectonic plate overrides and forces down the Pacific plate.*

- **New Zealand** is only a few thousand miles from the Antarctic Circle at its southern tip. As a result New Zealand has mild summers and cold winters.

- **Oceania's highest peaks** are Puncak Jaya, at 16,026 ft, and Mount Wilhelm at 14,790 ft, both on New Guinea.

- **The Great Barrier Reef** is the world's largest living structure, at 1,400 mi long. It is the only structure built by animals that is visible from space.

- **Although Australia sits** on the Indian-Australian plate, which is moving very slowly away from Antarctica, New Zealand sits astride the boundary with the Pacific plate.

- **Oceania is usually regarded** as having three sets of islands or cultural regions.

- **Micronesia is the region to the north** of New Guinea and east of the Philippines.

- **Melanesia includes** New Guinea and islands to the east, to Vanuata and Fiji.

- **Polynesia is to the east** of these two regions, from Hawaii in the north to New Zealand in the south, eastward over halfway across the Pacific to Easter Island.

Antarctica

- **Antarctica is the ice-covered continent** at the South Pole. It covers an area of 5,400,000 sq mi.

- **It is the coldest place on Earth**. Even in summer, temperatures rarely climb above −13°F. July 21, 1983, the air at the Vostok science station plunged to −128.5°F .

- **Antarctica is one of the driest places** on Earth, receiving barely any rain or snow. It is also very windy.

- **Until about 80 million years ago** Antarctica was joined to Australia.

- **Glaciers began to form** in Antarctica 38 million years ago, and grew rapidly from 13 million years ago. For the past five million years, 98 percent of the continent has been covered in ice.

▲ *Penguins are among the few creatures that can survive the cold of Antarctica all year round.*

● **The Antarctic ice cap** contains 70 percent of the world's fresh water.

● **The ice cap is thickest**—up to 15,700 ft deep—in deep sea basins far below the surface. Here it is thick enough to bury the Alps.

● **Antarctica is mountainous**. Its highest point is the Vinson Massif, at 16,863 ft.

● **The magnetic South Pole**—the pole to which a compass needle points—moves 5 mi a year.

● **Fossils of tropical plants** and reptiles show that Antarctica was at one time much warmer.

North America

- **The world's third largest continent** is North America. It has an area of 9,530,000 sq mi.

- **North America's long north side** is bound by the icy Arctic Ocean, and its short southeast side by the Gulf of Mexico.

- **The north lies inside** the Arctic Circle and is icebound for much of the year.

- **Death Valley**, in the southwestern desert in California and Nevada, holds the record as the hottest places on the Earth, at 134°F.

- **Mountain ranges** run down each side of North America—the ancient, worn-down Appalachians in the east and the younger, higher Rockies in the west.

▼ *A herd of bison moves among the billowing steam and spurting hot water of geysers, caused by underground "hot spot" rocks in Yellowstone National Park.*

▲ *Niagara Falls on the border between the U.S.A. and Canada pours more than 106,000 cu ft of water per second at peak flow.*

- **In between the mountains** lie vast interior plains. These are based on very old rocks, the oldest of which are in the Canadian Shield in the north.

- **Most of North America** sits on the North American tectonic plate. There are three hot spots beneath it at Yellowstone, Anahim, and Raton, generating volcanic activity.

- **The Grand Canyon** is one of the world's most spectacular gorges. It is 277 mi long, and 6,000 ft deep in places.

- **The longest river** in North America is the Mississippi–Missouri, at 3,700 mi long.

- **The highest mountain** is Denali in Alaska, at 20,321 ft.

- **The Great Lakes** in northeastern North America contain one-fifth of the world's fresh water.

South America

- **The world's fourth largest** continent, South America has an area of 6,884,000 sq mi.

- **The Andes Mountains**, which run over 2,800 mi down the west side, are the world's longest mountain range.

- **The heart of South America** is the vast Amazon rain forest around the Amazon River and its tributaries.

▼ *The high Andes Mountains have many glaciers, such as Perito Moreno glacier in southern Argentina. The glacier is 19 mi long and covers more than 97 sq mi. It flows downward 5,900 ft, east into Lake Argentina.*

▲ The jaguar is South America's largest big cat. It is at home in the swampy regions of the Amazon and Pantanal, where it swims well to catch fish and turtles.

- **The southeast** is dominated by the huge grasslands of the Gran Chaco, the Pampas, and Patagonia.

- **No other continent** reaches so far south. South America extends to within 620 mi of the Antarctic Circle.

- **Three-quarters of South America** is in the tropics. In the high Andes are large zones of cool, temperate climate.

- **Quito, in Ecuador**, is called the "Land of Eternal Spring" because its temperature never drops below 46°F at night, and never climbs above 71°F during the day.

- **The highest volcanic peak** is Aconcagua, at 22,841 ft.

- **Eastern South America** was joined to western Africa until the Atlantic began to open up 90 million years ago.

- **The Pantanal** in southwest Brazil, Paraguay, and Bolivia, is the world's biggest tropical wetlands, covering over 69, 500 sq mi in the wet season.

149

Major islands

- **The continents** of Eurasia and the Americas are technically islands as they are surrounded by ocean, but Australia is usually considered to be the largest island as well as the smallest continent.

- **Antarctica is actually** several islands joined by a single sheet of ice but it is sometimes considered to be one island as well as a continent.

▼ *Greenland is about four-fifths covered in the Greenland ice sheet, which formed at least half a million years ago. It is often crossed by airliners on commercial flights between northern North America and northern Europe.*

- **Greenland, part of Denmark**, is the largest island that is not also considered a continent. It has an area of 822,700 sq mi.

- **New Guinea** in the Pacific Ocean is part of Indonesia. It is the second largest island in area at 303, 360 sq mi.

- **Borneo**, part of the continent of Asia, is the third largest island at 288,800 sq mi.

- **Madagascar in the Indian Ocean**, is both an island and a country. At 226,900 sq mi it is the fourth largest island.

- **Baffin Island** lies to the northeast of Canada, of which it is part. At 195,900 sq mi it is the fifth largest island in the world.

- **The sixth largest island** at 182,800 sq mi is Sumatra, one of the Indonesian islands.

- **The seventh largest** is the Japanese island of Honshu. Its area is 87,180 sq mi, and it also has the largest population of the two main Japanese islands.

- **Victoria Island**, in the Northwest Territories and Nunavut of Canada, is eighth biggest, at 83,900 sq mi.

- **Great Britain** is the ninth largest island at 80,820 sq km. It is also the largest island in Europe.

- **Ellesmere Island**, also in Canada's Nunavut territory, is the tenth largest at 75,765 sq mi.

"World Ocean"

- **All the main oceans in the world**—the Arctic, Pacific, Atlantic, Indian, and Southern—are connected. For example, the Southern Ocean is at the southern ends of the Pacific, Atlantic, and Indian Oceans.

- **This means seas**, which are smaller but still huge areas around the edges of oceans, are also interconnected.

- **Together the salty waters** of oceans and seas occupy 71 percent, or over two-thirds, of the Earth's surface.

- **This interlinking** means that all seas and oceans can be regarded as one giant body of water, often called the "World Ocean."

- **The World Ocean** contains 97 percent of all the Earth's water—312 million cu miles.

- **The other 3 percent** of water on the planet is the freshwater in lakes, rivers, streams, and ponds.

- **The average depth** of the World Ocean is just over 13,000 ft.

- **The average temperature** of the World Ocean is only 39°F since the bulk of deep water is very cold.

- **Oceans are divided** by depth or vertically, and also across their areas, into zones.

- **The pelagic zones** are the open sea and ocean, away from coasts and also not near the seabed.

DID YOU KNOW?
The Northern Hemisphere is about three-fifths ocean, but the Southern Hemisphere is less than one-fifth ocean.

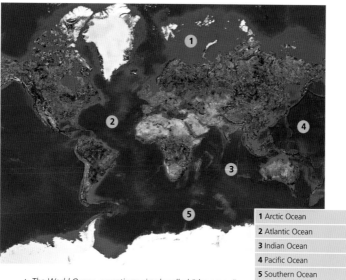

1	Arctic Ocean
2	Atlantic Ocean
3	Indian Ocean
4	Pacific Ocean
5	Southern Ocean

▲ The World Ocean, sometimes simply called "the ocean" or "the sea," covers twice as much of the planet as land does. The Southern Ocean has no land along its northern boundary, only other oceans, at latitude 60° South.

● **The benthic zones** are the deepest waters, near and on the bottom.

● **Oceans contain** many different habitats with different temperatures, salinity (saltiness), light levels, tides, and currents.

● **Currents of moving water** swirl around each ocean, carrying with them heat from the Sun, which affects the climate of the surrounding landmasses.

153

Atlantic Ocean

- **The Atlantic Ocean** is the world's second largest ocean, with an area of 41,000,000 sq mi. It covers about one-fifth of the world's surface.

- **At its widest point**, between Spain and Mexico, the Atlantic is 6,000 mi across.

- **The Atlantic was named** by the ancient Romans after the Atlas Mountains of North Africa. The Atlas were at the limit of the Romans' known world.

- **There are very few islands** in the middle of the Atlantic Ocean. Most lie close to the continents.

▼ *This computer model, built from sonar data, reveals the great ridge that winds along the floor of the Atlantic Ocean.*

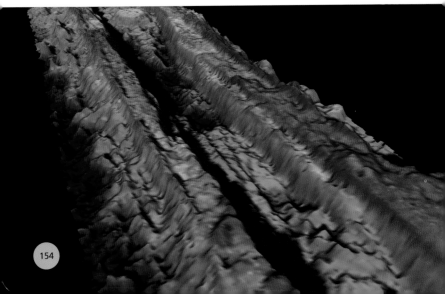

▲ *The nine main volcanic islands of the Azores are far out in the Atlantic, more than 800 mi west of Portugal in Europe. For its size, the Atlantic has far fewer islands than any other ocean.*

- **On average**, the Atlantic is about 11,100 ft deep.

- **The deepest point** in the Atlantic is the Puerto Rico Trench off Puerto Rico, which is 11,155 ft deep.

- **The Mid-Atlantic Ridge** is a great undersea ridge that splits the seabed in half. Here, the Atlantic is growing wider by about one inch every year.

- **Islands in the mid-Atlantic** are volcanoes that lie along the Mid-Atlantic Ridge, such as the Azores and Ascension Island.

- **The Sargasso Sea** is a huge area of water in the western Atlantic. It is famous for its floating seaweed.

- **The Atlantic** is a youngish ocean, less than 150 million years old.

Arctic Ocean

- **The smallest ocean** is the Arctic, at 5,424,700 sq mi.

- **It is also the shallowest** with an average depth of 3,400 ft.

- **Most of the Arctic Ocean** is permanently covered with a vast floating raft of sea ice.

- **Temperatures are low** all year round, averaging −22°F in winter and sometimes dropping to −94°F.

- **During the long winters,** which last more than four months, the Sun never rises above the horizon.

- **The Arctic gets its name** from *arctos*, the Greek word for "bear," because the Great Bear constellation is above the North Pole.

- **There are three kinds of sea ice** in the Arctic: Polar ice, pack ice, and fast ice.

- **Polar ice** is the raft of ice that never melts. It may be as thin as 6.5 ft in places in summer, but in winter it is up to 66 ft thick.

- **Pack ice forms** around the edge of the polar ice and only freezes completely in winter.

- **The ocean swell breaks** and crushes the pack ice into chunky blocks and fantastic ice sculptures.

- **Fast ice forms** in winter between pack ice and the land around the Arctic Ocean. It gets its name because it is held fast to the shore. It cannot move up and down with the ocean as the pack ice does.

▼ *Much of the Arctic Ocean freezes over each winter, then melts again in the summer.*

Indian Ocean

- **The Indian Ocean** is the third largest ocean. It covers one-fifth of the world's ocean area at 28,379,000 sq mi.

- **The average depth** of the Indian Ocean is 12,700 ft.

- **The deepest point** is the Java Trench off Java, in Indonesia, which is 24,442 ft deep. It marks the line where the Australian plate is being subducted under the Eurasian plate.

- **The Indian Ocean** is 6,200 mi across at its widest point, between Africa and Australia.

- **Scientists have calculated** that the Indian Ocean began to form about 200 million years ago when Australia broke away from Africa, followed by India.

- **Every year** the Indian Ocean gets 8 in wider.

- **The Indian Ocean is scattered** with thousands of tropical islands such as the Seychelles and Maldives.

- **The Maldives** are so low-lying that they may be swamped if global warming melts more polar ice.

- **Unlike other oceans**, currents in the Indian Ocean change course twice a year. They are blown by monsoon winds toward Africa in winter, and then in the other direction toward India in summer.

- **At the Rodrigues Triple Point** in the southern Indian Ocean, three tectonic plates—African, Indian-Australian, and Antarctic— all move away from each other due to seafloor spreading.

▼ The Maldives are some of many island groups in the Indian Ocean formed by coral on top of undersea volcanoes.

DID YOU KNOW?
The Persian Gulf is the warmest sea in the world whilst the Red Sea is the saltiest sea.

Pacific Ocean

◀ Easter Island, in the remote Central Pacific, is easily recognized by the great stone head statues called moai, made between 1200 and 1500.

- **The world's largest ocean** is the Pacific. It is twice as large as the Atlantic and covers over one-third of the Earth, with an area of 64,479,000 sq mi.

- **It is over 15,000 mi across** from the Malay Peninsula to Panama —more than halfway round the world.

- **The word "pacific" means peaceful**. The ocean was was given its name by 16th-century Portuguese explorer Ferdinand Magellan (1480–1521) who was lucky enough to find gentle winds when he first sailed there.

- **Thousands of islands** are dotted within the Pacific Ocean. Some are the peaks of undersea volcanoes. Others are coral reefs on top of the peaks.

- **The Pacific has some of the greatest tides**, over 30 ft off Korea. Its smallest tide, just one foot, is on Midway Island in the middle of the Pacific.

- **On average**, the Pacific Ocean is 14,000 ft deep.

- **Around the rim** there are deep ocean trenches including the world's deepest, the Mariana Trench.

- **A huge undersea mountain range** called the East Pacific Rise stretches from Antarctica, then north to Mexico.

- **The floor of the Pacific** is spreading along the East Pacific Rise at the rate of 4–6 in per year.

- **There are more seamounts** (undersea mountains) in the Pacific than any other ocean.

Southern Ocean

- **The fourth largest**, and youngest, ocean is the Southern or Antarctic. Its boundaries with the Pacific, Atlantic, and Indian Oceans have been debated but are generally considered to be latitude 60° South.

- **It is a deep ocean**, between 13,000 and 16,000 ft over most of its area, and 23,700 ft at its deepest.

- **The Southern Ocean** stretches all the way around Antarctica, and has an area of 13,500,000 sq mi. It is the only ocean that goes all around the world.

- **In winter** over half the Southern Ocean is covered with ice and icebergs that break off the Antarctic ice sheet.

- **Sea ice** forms in round pieces called pancake ice.

▼ The Southern Ocean may be cold and remote but its waters teem with the fish these emperor penguins feed on.

● **The East Wind Drift** is a current that flows anticlockwise around Antarctica close to the coast.

● **Further out from the coast** of Antarctica, the Antarctic Circumpolar Current flows in the opposite direction—clockwise from west to east.

● **The Antarctic Circumpolar Current** carries more water than any other current in the world.

● **The latitudes just below** 60° South are called the "Screaming Sixties," due to the fierce westerly winds that blow unobstructed for weeks at a time, all around the planet.

● **In this region,** sustained winds exceeding 93 mph whip up waves more than 50 ft high.

▼ *Blocks and chunks of ice continually break off into the Southern Ocean from the ice shelves and glaciers spreading outward from Antarctica. Some of the largest are bigger than small countries.*

163

Seas

- **Seas are small oceans**, enclosed or partly enclosed by land.

- **There are no major currents** flowing through seas and they are shallower than oceans.

- **In the Mediterranean** and other seas, tides can set up a seiche—a standing wave that sloshes back and forth like a ripple running up and down a bath.

- **If the natural wave cycle** of a seiche is different from the ocean tides, the tides are canceled out.

- **If the natural wave cycle** of a seiche is similar to the ocean tides, the tides are magnified.

- **About 6 million years ago**, the Mediterranean was cut off from the Atlantic and water evaporated, leaving a desert with just a few very salty lakes.

- **Then about 5.3 million years ago**, in the Zanclean flood, an earthquake allowed the Atlantic to gush through the land where the Straits of Gibraltar are now, filling the Mediterranean again.

DID YOU KNOW?
The Sea of Marmara, in Turkey between the Aegean and Black Seas, has an area of just 4,363 sq mi.

- **Warm seas such as the Mediterranean** lose much more water by evaporation than they gain from rivers. So a current of water flows in steadily from the Atlantic Ocean.

- **Warm seas lose** so much water by evaporation that they are usually saltier than the open ocean.

BLACK SEA

▶ The Mediterranean
is the world's largest
sea, covering about
965,000 sq mi.

MEDITERRANEAN SEA

165

Tides

▼ As the tide goes out here in St. Ives in southwest England (and other places in the world) it is flowing in as a high tide at other locations around the globe.

DID YOU KNOW?

On very gently sloping shores like mudflats, the rising tide water's edge moves faster than a person can run.

- **Tides are the way** the sea rises and falls every 12 hours. When the tide is flowing, it is rising. When the tide is ebbing, it is falling.

- **The pull of gravity** between the Earth, Moon, and Sun causes tides. Ocean waters flow freely over the Earth to create two tidal bulges (high tides) of water. One bulge is below the Moon, the other is on the opposite side of the Earth.

- **As the Earth turns**, and the Moon orbits, the two bulges stay in line with the Moon—and so they move around the Earth to create two high tides a day. But the Moon's slight movement means they occur every 12.5 hours, not 12 hours.

- **The continents get in the way** of the movement of tides, and as a result their timing and height varies. In the open, tides rise 3 ft or so, but in enclosed spaces such as the Bay of Fundy, Canada, they rise over 50 ft.

- **The Sun is much farther away** than the Moon, but it is so huge that its gravity affects the tides.

- **The Moon and the Sun** line up at a Full and a New Moon, creating high spring tides twice a month. (They have nothing to do with the season of spring.)

- **When the Moon and the Sun pull** at right angles at Half Moon, they cause neap tides, which are lower than normal tides.

- **The mutual pull** of the Moon's and the Earth's gravity also stretches the Earth slightly into an egg shape.

167

Waves

- **Waves are formed** when wind blows across the sea and whips the surface into ripples.

- **Water particles** are dragged a short distance by the friction between air and water, which is known as wind stress.

- **If the wind continues to blow** long and strong enough in the same direction, moving particles may build up into a ridge of water. At first this is a ripple, then a wave.

- **Waves seem to move** but the water in them stays in the same place, rolling around like rollers on a conveyor belt.

- **The size of a wave** depends on the strength of the wind and how far it blows over the water (the fetch).

- **If the fetch is short**, the waves may simply be a chaotic, choppy sea. If the fetch is long, they may develop into a series of rolling waves called a swell.

- **The biggest waves** occur south of South Africa.

- **The tallest breakers** rise to more than 98 ft before crashing onto the shore.

- **When waves move** into shallow water, the rolling at the base is impeded by the seabed. The water at the surface piles up, then spills over in a breaker.

DID YOU KNOW?

A wave over 130 ft high was recorded by the USS Ramapo in the Pacific in 1933.

▼ Waves break as the top spills over in shallow water.

Beaches

- **Beaches are slopes** of silt, sand, shingle, or pebbles along the edge of an ocean, sea, or lake.

- **Some beaches** are made entirely of broken coral or shells.

- **On a steep beach**, the backwash after each wave is strong. It washes material down the beach and so makes the beach slope more gently.

- **On a gently sloping beach**, each wave runs in powerfully and falls back gently. Material gets washed up the beach, making it steeper.

- **The slope of a beach** matches the waves, so the slope is often gentler in winter when the waves are stronger.

- **A storm beach** is a ridge of gravel and pebbles flung high above the normal high-tide mark during a storm.

- **At the top of each beach** a ridge, or berm, is often left at high tide mark.

- **Beach cusps** are tiny bays in the sand that are scooped out along the beach when waves strike it at an angle..

- **Many scientists believe** that beaches are only a temporary phenomena caused by the changes in sea levels after the last Ice Age.

- **The world's longest natural beach** is Cox's Bazaar, Bangladesh, at 77 mi.

DID YOU KNOW?

In 2007, the tallest sand castle, at Myrtle Beach in South Carolina, U.S.A, was recorded at 50 ft.

▼ *The white beach at Cancun, Mexico, is made from the fragments of coral reefs.*

Rocky coasts

- **Coastlines are changing** all the time as waves roll in and out and tides rise and fall. Over longer periods, coastlines are reshaped by the action of waves and the corrosion of salty water.

- **On exposed coasts** where waves strike the high rocks, they undercut the slope to create steep cliffs and headlands. Waves may penetrate into the cliff to form sea caves or arches. When a sea arch collapses, it leaves behind tall pillars called stacks.

- **Waves reshape rocks** by pounding them with a huge weight of water. Waves also force air into cracks in the rocks, which can split them open.

▼ Along coastlines, the sea wears away land, but it also deposits particles that settle and gradually form new land. This creates a variety of natural features along coasts.

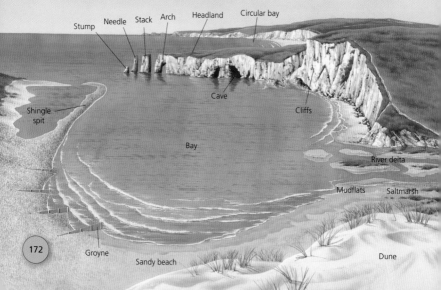

Stump · Needle · Stack · Arch · Headland · Circular bay

Cave

Shingle spit

Cliffs

Bay

River delta

Mudflats · Saltmarsh

Groyne

Sandy beach

Dune

▲ On many coastlines, waves carve away huge amounts of rock to leave isolated stacks like the Apostles on Australia's southern coast.

▲ Where hills of tough chalk rock meet the sea, the base is often worn away by waves to form steep cliffs.

- **The erosive power of waves** is focused on a narrow band at wave height. As cliffs retreat, the waves slice away a broad shelf of rock called a wave-cut platform.

- **When waves hit a coast** at an angle, they fall back down at a right angle to that first angle. This motion moves material along in a zigzag fashion. This is called longshore drift.

- **Longshore drift** can wash sand or shingle out across bays and estuaries to create long, narrow areas called spits or bars.

- **Bays are broad indents** in the coast with a headland on each side.

- **A cove is a small bay**. A bight is a huge bay, such as the Great Australian Bight. A gulf is a long, narrow bight.

- **The world's largest bay** by area is the Bay of Bengal, India, which is 850,000 sq mi. Hudson Bay, Canada, has the longest shoreline at 7,600 mi.

Coral reefs

- **Coral reefs** are made from calcium minerals, especially calcium carbonate, or chalk, a type of rock that is made by tiny creatures called polyps that live there in colonies.

- **Coral polyps** are related to jellyfish. Each one has tentacles around its mouth to catch food particles drifting by, and each one makes a small, hard mineral cup to live in.

- **Coral reefs** mostly grow in shallow water along continental shelves or around extinct marine volcanoes or seamounts.

- **About 0.1 percent** of the ocean surface is covered by coral reefs —about 110,000 sq mi.

▼ *Almost 1,200 mi of coral reefs fringe the shores of the Red Sea, between Africa and Arabia. They are home to more than 1,200 species of fish.*

◄ *Most coral polyps are smaller than your fingertip, but some grow to 4 in tall. They are active mainly at night when, like their cousins, the jellyfish and anemones, they wave their tentacles to catch food.*

- **Most corals form** in the warm tropical waters of the Pacific and Indian Oceans.

- **The Great Barrier Reef**, off the eastern coast of Australia, is the largest reef system in the world at 1,400 mi long.

- **The second largest single reef** is the Mesoamerican Barrier Reef system to the east of Central America, from Yucatan to Honduras, at 620 mi long.

- **The New Caledonia Barrier Reef system** is a double reef and it is 900 mi long, second only to the Great Barrier Reef.

- **Coral reefs are among** the most diverse ecosystems in the world. More than 25 percent of all the animals in the sea live on coral reefs.

- **Coral polyps** are very sensitive to water temperature. They grow best at 78–80°F and global warming of the oceans kills or "bleaches" them.

- **They are also sensitive to pollution**, since their chalky cases dissolve where chemicals make the water acidic.

Icebergs

- **Icebergs are big lumps** of floating ice that calve, or break off, from the ends of glaciers or polar ice caps.

- **The calving of icebergs** occurs mostly during the summer when the warm conditions partially melt the ice.

- **Arctic icebergs** vary from car-sized blocks, to those the size of mansions. The biggest, which was 7 mi long, was spotted off Baffin Island in 1882.

- **The Petterman and Jungersen glaciers** in northern Greenland form big table-shaped icebergs called ice islands. They are like the icebergs found in Antarctica.

- **Antarctic icebergs** are much bigger than Arctic ones. One of the biggest, which was 186 mi long, was spotted in 1956 by the icebreaker USS *Glacier*.

- **Another 186-mi iceberg**, B–15, came from the Ross Ice Shelf in Antarctica in 2000.

- **The ice of Arctic icebergs** is 3,000–6,000 years old.

- **The International Ice Patrol** was set up in 1914 to monitor icebergs after the liner RMS *Titanic* sunk in 1912, when it hit an iceberg off Newfoundland.

▼ *Rocked by the waves, chunks of ice break off Antarctic glaciers and ice sheets to form fantastically shaped icebergs.*

Ocean deeps

- **The oceans** are over 13,000 ft deep on average.

- **Along the edge** of the ocean is a ledge of land called the continental shelf. The average water depth here is 400–700 ft.

- **At the edge** of the continental shelf the seabed plunges thousands of feet steeply down the continental slope.

- **The gently sloping foot** of the continental slope is called the continental rise.

- **Beyond the continental rise** the ocean floor stretches out in vast plains called abyssal plains. They lie about 13,000–20,000 ft below the water's surface.

- **The abyssal plain** is covered in a thick slime called ooze. It is made from volcanic ash, meteor dust, and the remains of dead sea creatures that sink down as "marine snow."

- **The abyssal plain is dotted** with huge mountains, thousands of feet high, called seamounts.

- **Flat-topped seamounts** are called guyots. They may be volcanoes that once projected above the surface.

- **The deepest places** in the ocean floor are ocean trenches. These are made along subduction zones, where oceanic tectonic plates are driven down into the mantle by continental plates. The deepest is the Mariana Trench in the Pacific.

DID YOU KNOW?

Abyssal plains cover almost half of the ocean floor, so they have about the same area as all the land on Earth.

▼ *The ocean floor gets deeper in distinct zones. In most oceans the abyssal plain stretches flat and wide for thousands of miles.*

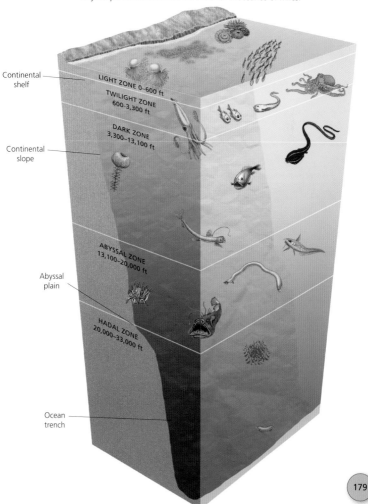

Continental shelf

LIGHT ZONE 0–600 ft

TWILIGHT ZONE 600–3,300 ft

DARK ZONE 3,300–13,100 ft

Continental slope

ABYSSAL ZONE 13,100–20,000 ft

Abyssal plain

HADAL ZONE 20,000–33,000 ft

Ocean trench

179

Black smokers

- **Billowing black fumes** of hot gases, particles, and scorching water, black smokers are natural chimneys on the seabed.

- **The technical name** for black smokers is hydrothermal vents. They are volcanic features.

- **Black smokers form** along mid-ocean ridges where the tectonic plates are moving apart.

- **When seawater seeps** through cracks in the seafloor, black smokers may form. The water is heated by volcanic magma, and it dissolves the minerals in the rock.

- **Once the water** is superheated, it spews from the vents in scalding, mineral-rich black plumes.

● **The plume cools rapidly** in the cold sea, leaving behind thick deposits of sulfur, iron, zinc, and copper in chimneylike vents.

● **The tallest vents** can exceed 160 ft in height.

● **Water jetting** from black smokers can reach a temperature of 12,000°F.

● **Smokers are home to a community** of organisms that thrive in the scalding waters and toxic chemicals. The organisms include giant clams, blind crabs, eelpout fish, and giant tubeworms.

DID YOU KNOW?
Each drop of sea water in the oceans circulates through a black smoker every ten million years.

◄ Black smokers belch streams of superheated water, hot gases, and mineral particles into the cold, inky ocean depths.

Surface ocean currents

- **Ocean surface currents** are like giant rivers many miles wide, on average 330 ft deep and flowing at 9 mph.

- **The major currents** are split on either side of the Equator into giant rings called gyres.

- **In the Northern Hemisphere** the gyres flow clockwise. In the south they flow anticlockwise.

▼ *The warm Gulf Stream current flowing across the Atlantic (light pink) helps to keep the climate of northwest Europe mild.*

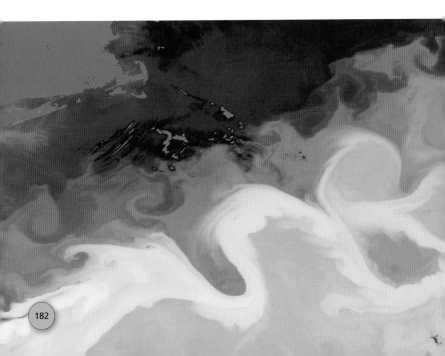

- **Ocean currents** are driven by a combination of winds and the Earth's rotation.

- **Near the Equator**, water is driven by easterly winds to make westward-flowing equatorial currents.

- **When equatorial currents** reach continents, the Earth's rotation deflects them poleward as warm currents.

- **As warm currents flow polewards**, westerly winds drive them east back across the oceans. When the currents reach the far side, they begin to flow toward the Equator along the west coasts of continents as cool currents.

- **The North Atlantic Drift**, the northern part of the Gulf Stream, brings warm water from the Caribbean to southwest England, so it is warm enough to grow palm trees, yet it is as far north as Newfoundland.

- **By drying out the air**, cool currents can create deserts, such as California's Baja and Chile's Atacama deserts.

Deep ocean currents

- **Ocean surface currents** affect only the top 330 ft or so of the ocean. Deep currents involve the whole ocean.

- **Deep currents** are set in motion by differences in the density of sea water. Most move only a few feet a day.

- **Most deep currents** are called thermohaline circulations because they depend on the water's temperature ("thermo") and salt content ("haline").

- **If sea water is cold** and salty, it is dense and sinks.

- **Typically, dense water forms** in the polar regions. Here the water is cold and weighed down by salt left behind when sea ice forms.

- **Dense polar water** sinks and spreads out toward the Equator far below the surface.

- **Oceanographers** call dense water that sinks and starts deep ocean currents "deep water."

- **In the Northern Hemisphere** the main area for the formation of deep water is the North Atlantic.

DID YOU KNOW?
Antarctic deep water is very cold, at 28.6°C, but its salt and movement prevent it freezing solid.

- **Dense salty water** from the Mediterranean sinks quickly—3.3 ft per second—through the Straits of Gibraltar to add to the North Atlantic deep water.

- **There are three** vertical levels in the ocean.

- **The epilimnion** is the surface waters warmed by sunlight, usually 300–1,000 ft deep.

- **The thermocline** is where the water becomes colder quickly with depth, usually between 1,600 and 3,300 ft.

- **The hypolimnion** is the vast bulk of deep, cold ocean water.

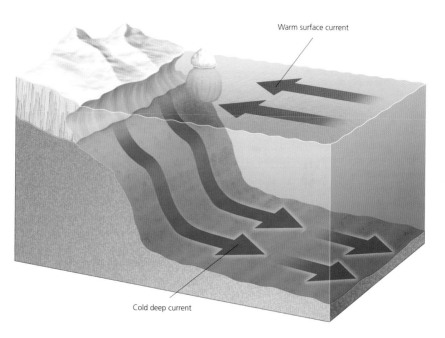

Warm surface current

Cold deep current

▲ *Deep water circulations begin in the polar regions where cold, dense water sinks, laden with salt left behind as ice forms.*

185

Ocean trenches

- **Trenches—the steep, deep valleys** at the bottom of oceans—are found where tectonic plates are sliding past each other or where one slides under another, called subduction.

- **These zones** are often marked by lines of volcanic islands or seamounts.

- **Gravity measurements** over a trench show that a "downwelling," where higher density rocks are sinking beneath lighter density rocks, is occurring deep beneath the Earth's crust.

- **The existence of trenches** was unknown before exploration took place to find suitable seabed to lay transatlantic telegraph cables at the end of the 19th century.

- **More than 20 deep sea trenches** have been discovered in the world's oceans: 18 in the Pacific Ocean, three in the Atlantic and one in the Indian Ocean.

- **The deepest** is the Mariana Trench.

- **The Tonga Trench** in the Pacific Ocean, at 35,702 ft, is the second deepest trench in the world.

- **The Philippine Trench**, also in the Pacific Ocean, is 34,580 ft.

- **The Kuril-Kamchatka Trench** also in the Pacific Ocean is 34,586 ft.

- **The next deepest** are the Kermadec Trench in Pacific Ocean at 32,962 ft, and the Izu-Ogasawara Japan Trench in the Pacific Ocean at 31,181 ft.

● **The Peru-Chile**, or Atacama, Trench runs just offshore for
3,600 mi along the entire Pacific coast of South America. It is
5,011 ft deep, making it the ninth deepest of all the trenches.

▼ *Most dives into trenches are by unmanned craft called ROVs (remotely operated underwater vehicles). They carry cameras, sensors for pressure, temperature, and salinity, robot arm sample-grabbers, and many other devices. This photograph shows the craft* Jiaolong *taking samples almost 4,350 ft down in the Mariana Trench.*

Mariana Trench

- **The Mariana Trench** is in the Pacific Ocean, east of the Mariana Islands and south of Japan.

- **It is 1,398 ft long** and averages 44 mi wide.

- **It is the deepest part** of the world's oceans—estimates vary but most show that it is around 36,000 ft at the lowest point.

- **This deepest point**, at the southern end of the trench, is called the Challenger Deep after the research vessel that discovered it in 1875.

- **The water pressure** at the bottom of Challenger Deep is 1,000 times greater than at the ocean's surface—about 9 tons per sq in.

- **The temperature** at the bottom of the trench is 33–39°F.

- **The Mariana Trench** is formed where the Pacific tectonic plate is being pushed, or subducted, under the Mariana Plate.

- **This subduction** has also created the volcanoes that have formed the nearby Mariana Islands.

- **The first manned expedition** to the bottom of Challenger Deep was by Jacques Piccard and Don Walsh in the bathyscaphe *Trieste* on January 23, 1960.

- **It took 4 hours 47 minutes** to dive from the surface to the bottom, and 3 hours 15 minutes to come back up—after just 20 minutes on the bottom.

- **The bottom of the trench** is covered with a deep ooze made up of the dead skeletons of tiny plantlike organisms (diatoms) that live in the sea water above.

▲ In 2012, movie director and producer James Cameron dived over 36,000 ft to the bottom of the Challenger Deep in the submersible Deepsea Challenger. The journey down took two and a half hours, with three hours on the bottom filming and taking samples, and less than two hours to return to the surface.

Puerto Rico Trench

▲ Ocean floors and depths are mapped by echosounding or sonar—sending out sound waves and measuring the echoes (reflections). This ship has a multibeam sonar and tows a submersible with sidescan sonar.

- **The Puerto Rico Trench** is an arc-shaped gorge 6,700 mi long that lies just to the north of Puerto Rico and the Virgin Islands.

- **It follows the edge** of the Caribbean Sea where it meets the Atlantic Ocean.

- **The deepest part** of the trench is the Milwaukee Depth, at about 27,500 ft. It was discovered using sonar by the U.S. navy ship of that name, in 1939.

- **The Milwaukee Depth** is also the deepest point in the Atlantic and the deepest in any ocean outside the Pacific Ocean.

- **The trench lies between** two tectonic plates, the North American plate and the Caribbean plate, which are sliding past each other at a rate of 0.8 in per year.

- **This sliding is not gradual** but occurs as a phenomenon known as fault-slip: the plate edges stick and then suddenly slip when pressure builds up.

- **This leads to** frequent powerful earthquakes in the area, which in turn cause major tsunamis.

- **There have been many earthquakes** and tremors caused by this fault-slip system in the last 100 years. The subduction area has been quiet for a while, and scientists are expecting a major geological event at any time.

- **There was an earthquake** in the trench in 1787, with a magnitude of 8.1. Another one in 2014 measured 6.4.

- **Further south**, near the Lesser Antilles, there is a subduction zone where the Caribbean plate is sliding under the North American plate.

- **This subduction zone**, although small, causes the active volcanoes found in the southeast Caribbean Sea.

Mountains and canyons

Mountain systems

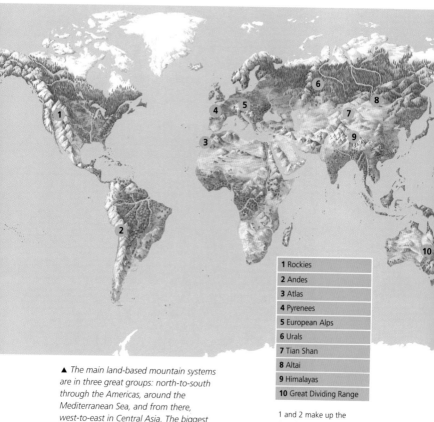

1	Rockies
2	Andes
3	Atlas
4	Pyrenees
5	European Alps
6	Urals
7	Tian Shan
8	Altai
9	Himalayas
10	Great Dividing Range

▲ The main land-based mountain systems are in three great groups: north-to-south through the Americas, around the Mediterranean Sea, and from there, west-to-east in Central Asia. The biggest system in Oceania is Australia's Great Dividing Range.

1 and 2 make up the American Cordillera.
7, 8, and 9 make up the Pamir-Himalayas.

- **Mountain systems** are groups or ranges of mountains that have been formed by the same geological events.

- **These systems often include** ranges and sub-ranges within them. Chains of mountain ranges are also sometimes called cordilleras.

- **The Alpine Belt** is a system that sweeps from the Alps of Europe, through the Himalayas to the mountains of Indonesia.

- **The system of mountains** called the Arctic Cordillera passes through northeast Canada, from Ellesmere Island south across Baffin Island to the northern tip of Newfoundland.

- **The longest unbroken mountain system** in the world is the mid-ocean ridge system, which is more than 37,200 mi long.

- **The Pacific Ring of Fire** is a very active mountain-building area all around the Pacific tectonic plate, called the Circum-Pacific Mountain System.

- **This system** can be said to include the America Cordillera, the Aleutians, Kamchatka, Japan, Taiwan, the Philippines, Papua New Guinea, and New Zealand.

- **The Andes** alone is sometimes said to be the longest mountain system in the world at 4,350 mi.

- **The Pacific Mountain System** stretches 4,500 mi from British Columbia in Canada down the western North American coast to Mexico.

- **The American Cordillera** mountain system starts in Alaska and passes down the western coast of North and South America to the southern tip of Chile.

Himalayas

- **The Himalayas** is a mountain range that lies between the subcontinent of India and Tibet, part of China. The mountains lie across or touch four other countries: Nepal, Bhutan, Afghanistan, and Pakistan.

- **"Hima" means "snow"** in Sanskrit, *alaya* means "dwelling," so himalaya means "home of the snow."

- **The Himalayas are covered** by more snow and ice than anywhere else in the world apart from the polar ice caps on Antarctica and the Arctic.

- **The 100 highest mountain peaks** in the world are in the Himalayas. They are all more than 23,000 ft tall.

▼ The Himalayas, here shown with Everest in the center, are known as the "roof of the world." The whole range is about 1,550 mi long.

- **The highest** of them all is Mount Everest at 29,028 ft.

- **Three of the most important rivers** in the world begin in the Himalayan mountains: the Ganges, Indus, and Brahmaputra. Combined they supply fresh water to one-fifth of all the people in the world.

- **The mountains** that make up the Himalayas are among the youngest in the world.

- **They began to form** about 70 million years ago when the Indo-Australian tectonic plate pushed against the Eurasian plate, finally closing up the Tethys Ocean that previously lay between them.

- **The mountains are still rising** at about 5 mm per year, as the Indo-Australian plate, which is moving northward at a rate of 2.7 in per year, pushes under the Eurasian plate.

- **Because of all the ice**, snow, meltwater, streams, and rivers, the mountains are eroding at a rate of 2–12 mm per year.

DID YOU KNOW?

The Tibetan Plateau is the world's highest. It covers 965,000 sq mi at an average height of 14,700 ft.

Andes

● **It is believed the word "Andes"** comes from the Quechua (South American peoples) word *antisuyu*, which was the "east region" of the Inca Empire.

▼ *Many Andes peaks are always covered by ice and snow. Yet animals survive here, such as these Andean condors, soaring away from their roosting ledge to scan the cliffs and valleys for food.*

- **Running along the west coast** of South America, the Andes is the longest continuous mountain range in the world at 4,350 mi, and between 124 and 435 mi wide.

- **The range passes through** seven South American countries: Venezuela, Colombia, Ecuador, Peru, Bolivia, Chile, and Argentina.

- **With an average height** of 13,120ft, the Andes is the highest mountain range outside of Asia.

- **Aconcagua in Argentina** is the highest of the Andes mountains, with its peak 22,841 ft above sea level.

- **The Andes lie along the edge** of the Pacific Ring of Fire, a circle of volcanic activity caused by the movement of tectonic plates around the edges of the Pacific Ocean.

- **The highest volcanoes** in the world are found in the Andes mountains. Tallest is Ojos del Salado at 22,614 ft.

- **Those mountains** that are not volcanoes were formed by uplifting, faulting, and folding as the Nazca and Antarctic plates were forced under the South American plate.

- **There is still geological activity** in the area and since subduction continues, many of the volcanoes in the Andes are still active.

- **Because of their long and active** geological history, the Andes mountains have rich deposits of resources such as gold, silver, copper, tin, iron, and nitrates, and they are the world's largest sources of lithium.

The Rockies

● **The Rocky Mountains**, or just "Rockies," are the main mountain range in North America.

● **The range is 3,000 mi long** and between 68 and 300 mi wide, passing from British Columbia in Canada to New Mexico in the U.S.A.

● **The Rockies** have undergone bouts of mountain-building for hundreds of millions of years. Some of the rocks were formed in Precambrian times nearly 2,000 million years ago.

● **The current range of mountains**, however, is mostly quite "young," forming 80–50 million years ago, when various tectonic plates pushed under the North American plate.

▶ At 14,110 ft tall, Pikes Peak in the eastern Rockies of Colorado is visible for many miles across the Great Plains. It became a symbol for pioneers approaching the mountains looking for gold in the 19th century.

- **The highest peak** is Mount Elbert in Colorado, U.S.A., which is 13,123 ft tall.

- **The Rockies** are part of a longer chain of mountain ranges called the American Cordillera that runs south from North America through Central America and South America to Antarctica.

- **Mineral resources** recovered from the Rocky Mountains include copper, gold, lead, molybdenum, silver, tungsten, and zinc. Coal, gas, and oil are also mined.

- **The Continental Divide** of North America runs through the Rocky Mountains. This is the source of some rivers that flow west to the Pacific and others that flow east to the Atlantic.

- **Rivers flowing west** include the Peace River, Columbia River, and Colorado River.

- **Rivers flowing east** include the Rio Grande, the Missouri, and the Platte River.

Urals

- **The Ural Mountains** are a range that runs for 1,550 mi in western Russia, from the Arctic Ocean south to Kazakhstan.

- **The range forms** a natural barrier between Europe and Asia.

▼ *Coal, other minerals, metals such as iron and copper, and many gems—here in an immense diamond mine—from the Urals supply much of Russian industry.*

DID YOU KNOW?

About 300 million years ago the Ural Ocean was squeezed shut by plate movements to form the Ural Mountains.

- **There is a legend** in Bashkortostan about a hero called "Ural" who was buried under a pile of rocks after he died to save his people. The rock-pile became the mountains.

- **Some 50 valuable deposits** are found here such as gold, iron, copper, mica, gemstones, coal, oil, and gas. Their exploitation led to settlements in desolate places and eventually to Russia's industrial development.

- **At 6,217 ft**, Mount Narodnaya is the highest mountain in the range.

- **The Urals** are among the world's oldest mountains, being formed up to 300 million years ago, when the ancient continent of Eurasia pushed against a continent called Kazakhstania.

- **The Ural and Kama Rivers** run south and east to the Caspian Sea. The Pechora River flows west and north to the Arctic Ocean.

- **The rivers** of the Ural Mountains are frozen solid for at least half of every year.

- **There are many deep lakes** in the mountains, with the deepest, Lake Bolshoye Shchuchye, at 446 ft.

- **The lakes and rivers** are fed by up to 3.3 ft of precipitation, mainly as snow, per year in the northern Urals.

European Alps

- **The Alps** stretch across Europe for 745 mi, from the Adriatic Sea in the east to the Mediterranean Sea in the west.

- **The range passes through** France, Italy, Switzerland, Austria, and Slovenia.

- **About 300 million years ago**, the mountains began to form as the African and Eurasian tectonic plates pushed together, crumpling the rock.

- **The mountains formed** from marine sedimentary rocks from beneath the ancient Tethys Ocean, which were uplifted causing "fold-and-fault" mountains.

- **Mont Blanc** is the highest mountain, on the border between France and Italy. It is 15,780 ft in height.

- **There are nearly 100 other peaks** in the Alps higher than 13,000 ft.

- **A frozen body** was found in the Alps in 1991. Known as Otzi the Iceman, he is 5,000 years old and has provided much fascinating information about life at that time.

▶ Otzi was preserved at a height of 10,500 ft with his clothes, weapons, tools, and even medicinal herbs.

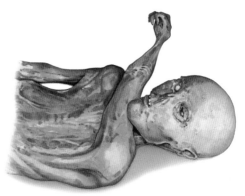

▲ *More than 100 million people visit the Alps yearly to ski, snowboard, climb, and walk in the fabulous scenery.*

- **The Alps are popular** with mountaineers. Mont Blanc was first climbed in 1788 while most of the other peaks ove 13,000 ft were scaled by the 1850s.

- **The last of the more difficult "north faces,"** that of the Eiger, was finally scaled in 1938.

- **People have quarried** and mined the Alps for thousands of years. The rocks are rich in copper, gold, iron, and crystals such as amethyst and quartz.

- **There is a laboratory** inside one of the Alps glaciers where scientists study the movement of the ice and the erosion it causes.

Great Dividing Range

- **Also called the Eastern Highlands**, the Great Dividing range is the main mountain range in Australia, running south from the northern tip of Queensland, through New South Wales to Victoria.

- **At more than 2,500 mi** it is the third longest mountain range, on land, in the world.

- **The range was formed** by faulting and folding 300 million years ago when the Australian plate collided with a plate that carried what is now New Zealand and part of South America.

- **Most of the mountains** formed in that event have completely eroded away, and those that are left are fairly small by world standards and are easy to climb.

- **Australia's highest mountain**, Mount Kosciuszko, on the Great Dividing Range, is 7,309 ft in height.

- **The mountains influence** the climate of the whole continent. Water vapor is brought to the east coast from the Pacific Ocean by the trade winds, and the mountains cause the water to fall as rain or snow on the eastern coast.

- **Lands in the rain shadow**, west of the mountains, get very little rainfall and are therefore very arid.

- **The mountains are called** the Great Dividing Range for a reason. They divide the rivers into those that flow from the mountains east to the Pacific Ocean, such as the Snowy River, and those that flow west toward the central lowlands, such as the Murray-Darling River system.

- **The first people** to settle in the ranges were the Australian Aboriginal tribes who decorated the caves and made paths across the mountain passes.

- **The range formed** a significant barrier to European settlers until English pioneers Gregory Blaxland, William Lawson, and William Charles Wentworth found a route through the Blue Mountains, part of the range, in 1813.

▼ *Hydroelectric dams in the Snowy Mountains, at the southern end of the Great Dividing Range, provide power to cities such as Canberra, Melbourne, and Sydney.*

Atlas

- **The Atlas Mountains** form a chain that runs for 1,550 mi across the northwest corner of the continent of Africa.

- **The chain covers an area** of 299,360 sq mi and passes through three countries—Algeria, Morocco, and Tunisia.

- **Toubkal in Morocco** is the highest mountain, and it rises 13,664 ft above sea level.

- **The Atlas Mountains** are bounded by the Atlantic Ocean in the northwest, the Mediterranean Sea in the north, and the Sahara in the south.

- **The range was formed** during three episodes of mountain building from Precambrian rocks, over billions of years.

- **The first episode** was in the Palaeozoic Era when Africa collided with America.

- **Episode two** was during the Mesozoic Era when Africa and America began to drift apart as new crust formed between them.

- **The third phase** occurred in the Cenozoic Era, between 66 and 1.8 million years ago, when Europe collided with Africa and pushed up the rocks to form the present mountains.

- **Mineral resources** such as iron, lead, copper, silver, mercury, coal, and gas are mined in the Atlas Mountains.

- **The whole range** consists of three smaller ranges: the Middle Atlas in the north, the High Atlas, and the Anti Atlas in the south.

▼ *Mount Toubkal is covered with snow and ice in winter but becomes parched and bare in the summer heat.*

Transantarctic Mountains

- **The Transantarctic Mountains** form a broken line stretching across the continent of Antarctica from the Weddell Sea to the Ross Sea.

▼ *Since Antarctica is a desert, with very little rain or snow, the main weathering forces on the Transantarctic Mountains are the sun's heat, bitter cold, and frost and ice wedging.*

- **They divide** the continent into East and West Antarctica.

- **The mountains stretch** for 2,175 mi —one of the longest mountain ranges on Earth.

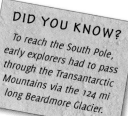

DID YOU KNOW?

To reach the South Pole, early explorers had to pass through the Transantarctic Mountains via the 124 mi long Beardmore Glacier.

- **While mountain peaks** in other parts of the world are often covered by ice and snow, in Antarctica some of the mountain peaks, and even valleys between them, are the only parts of the continent not covered by ice.

- **Many peaks rise** to over 13,000 ft above sea level. Highest is Mount Kirkpatrick at 14,855 ft.

- **Mountain peaks** that poke up through the ice are called nunataks.

- **The range began** about 65 million years ago by uplifting when the West Antarctic Rift formed, pushing rocky crust eastward.

- **The East Antarctic** is covered by a vast ice sheet that flows through gaps between the mountains as glaciers and then into the Ross Sea.

- **Temperatures fall** as low as −112°F and there is no fresh water. Only animals such as seals, penguins, and sea birds can live here, along with bacteria, lichens, algae, and fungi.

- **Coal reserves**, possibly the largest in the world, have been found under the mountains. But international treaties, and the desolate environment, prevent any mining activity.

Mid-Atlantic Ridge

- **The Mid-Atlantic Ridge** stretches 9,900 mi—the longest mountain range in the world, but most of it is submerged under the Atlantic Ocean.

- **It runs along the middle** of the Atlantic, following the shape of the Americas to the west, and Europe and Africa to the east.

- **In the North Atlantic**, the ridge marks the junction between the Eurasian tectonic plate and the North American plate.

- **In the South Atlantic** the ridge separates the African and South American plates.

- **There is a valley** or rift along the center of the ridge where hot liquid rock, or magma, wells up and solidifies to form new ocean crust, as seafloor spreading. This pushes the east and west plates apart at a rate of 0.7–1.5 in every year.

- **The newly formed rock** wrinkles either side of the ridge, into rows of parallel mountains and valleys.

- **The Mid-Atlantic Ridge** began to form about 180 million years ago when it split the ancient vast supercontinent of Pangaea.

- **It is part of a system** of ridges that stretches over 37,000 mi around the world beneath all the oceans.

- **Iceland is the tip** of one of the mountains in the chain. It is 6,919 ft from its highest point to the seafloor around.

- **Other islands** in the Atlantic Ocean that are ridge mountain tops are the Azores, Ascension Island, St. Helena, and Tristan de Cunha.

▲ *The Mid-Atlantic Ridge curves almost halfway around the planet from the Arctic Ocean almost to the continent of Antarctica.*

1	Iceland
2	Greenland
3	North Atlantic Ocean
4	Tore-Madeira Rise
5	Romanche Trench
6	South Atlantic Ocean

213

Grand Canyon

- **The Grand Canyon**, in Arizona, U.S.A., is a deep gorge 277 mi long and between 4 and 18 mi wide.

- **The Colorado River** runs along the bottom of the canyon, which is over 5,900 ft deep in places.

- **In fact the Grand Canyon** was formed by the Colorado River, which has been cutting its way through, or eroding, the rocks for around 17 million years.

▼ *Many of the rocks in the Grand Canyon, including sandstones and limestones, glow intense colors in the setting sun.*

- **The top layer of rock**, called Kaibab limestone, was formed about 240 million years ago.

- **As the river has cut deeper** it has exposed older and older layers of rocks. Those at the bottom are up to two billion years of age.

- **The geological record** is practically complete between these two dates except for a single gap, called an unconformity, between 500 million and 1,500 million years ago, when there was erosion but no deposition (new rock formed).

- **The rocks were mostly formed** as layers of sediment at the bottom of, or around the edges of, warm, shallow seas. They contain fossils of marine animals such as corals and sponges.

- **Some of the rock layers**, however, were laid down on land in swamps, rivers, or sand dunes. These contain fossils such as leaves, dragonfly wings, or the tracks of early land animals.

- **The Grand Canyon** is one of the natural wonders of the world. It is the most popular tourist site in Arizona and has about five million visitors every year.

- **The canyon cuts through** an area called the Colorado Plateau, a very stable area unaffected by significant mountain building for hundreds of millions of years.

Indus River Gorge

- **The Indus River** begins life as a mountain spring in Tibet called the Sengge, about 16,400 ft above sea level.

- **It becomes** the Indus River when it joins the Gar River.

- **Farther downstream**, near the Nanga Parbat mountain in Pakistan, it joins with yet another river, the Gilgit. Together they have eroded the deepest canyon in the world.

- **The river begins** so high in the mountains and falls so quickly that the churning water is a powerful erosion machine. It carries the eroded particles with it, which in turn wear away more rock.

- **Nanga Parbat**, 9 mi south of the river, is the ninth highest mountain in the world, at 26,660 ft.

- **The nearest mountain** to the north is Rakaposhi, 25,551 mi tall.

- **Experts have different sets** of measurements but the heights of these two mountains mean the Indus Gorge between them can be said to be 23,359 ft deep.

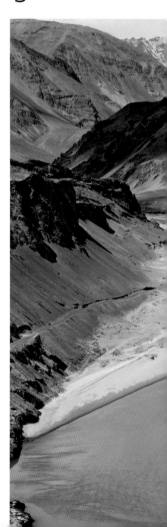

- **Other estimates** of the depth range from 14,700–17,000 ft.

- **The gorge** is between 12 and 16 mi wide.

- **Tributaries** of the rivers are fed by meltwater from the snow and glaciers high in the Himalayas.

- **Eroded material** is carried by the river all the way from the Indus Raiver Gorge into the sea, at its delta at Karachi, Pakistan.

- **Here, the material** has settled under the sea as the Indus Submarine Fan in the Arabian Sea—the second largest sediment body on Earth.

◀ *Where the Zanskar River joins the Indus, near Ladakh, India, the canyon has steep walls with almost sheer cliffs towering over 1,640 mi in height.*

Other canyons and gorges

- **Fish River Canyon**, Namibia, at 99 mi long, 16 mi wide and 1,804 ft deep, is usually regarded as the second largest canyon in the world in terms of size.

- **Kali Gandaki Gorge** in Nepal, 21,391 ft deep, can be seen as the second deepest gorge in the world.

- **Yarlung Tsangpo Gorge**, also in Nepal, is 313 mi long with an average depth of 18,277 ft, however it is not clearly one long continuous gorge.

- **Another gorge in Nepal**, the Kali Gandaki (or Andha Galchi), is thought by some to be the deepest. It forms a pass between the peaks of Dhaulagiri, 26,794 ft, and Annapurna, 26,545 ft.

- **Copper Canyon**—so-called because of its copper ore-green walls—in Chihuahua, Mexico, is deeper and longer than the Grand Canyon, but it is formed of six separate canyons with six different rivers.

◄ *The Three Gorges section of China's Yangtze River is both a world tourist destination with cruise ships and a busy commercial waterway for barges.*

- **In 2013** a canyon was discovered by radar under the ice sheet of Greenland. At 466 mi long, 2,624 ft deep, and 6.2 mi wide, it could technically be the biggest canyon in the world.

- **Provo Canyon** in Utah, U.S.A., is a giant rift between two mountain peaks in the Rockies Mountain ranges.

- **The road from Moldova to** Transylvania passes along the Bicaz Canyon, in northern Romania. It is 5 mi long and winds between a sheer drop on one side and vertical rocks above on the other, and is one of the most spectacular —and frightening—drives in the world.

- **The Yangtze River** (Chang Jiang) in China runs eastward through the scenic Three Gorges for 75 mi. The river was dammed in 2006 producing a reservoir 373 mi long.

Rivers through the ages

- **Rivers that are relatively short** and small today may actually be much older than the greatest and longest ones.

- **The Meuse River** that flows through France, Belgium, and the Netherlands is about 570 mi long, but it has probably been flowing for more than 360 million years.

- **The Finke River** in north-central Australia has been flowing on and off for more than 320 million years. At its longest it reaches 435 mi but in drought periods it shrinks because parts of its bed dry out.

- **Measurements of river lengths** vary according to where the river is said to begin. So some people say that Africa's Nile is the world's longest river; others say that South America's Amazon is longer.

- **Several sources of the Amazon** were discovered, making it longer and longer, the most recent in 2008.

- **The world's longest tributary** is the Madeira River, which flows into the Amazon. At 2,100 mi, it is the 18th longest river in its own right in the world.

- **The world's longest estuary** is that of the Ob in Russia, which is up to 50 mi wide and 550 mi long.

- **The Ob** is the biggest river to freeze solid in winter.

- **One of the shortest official rivers** is the North Fork Roe River in Montana, U.S.A., which is just 58 ft long.

- **The Kani Bil** in Iran, arising from a spring, flows for only 49 ft.

◄ In Novosibirsk, central Russia, the frozen Ob is useful even in winter. It hosts skate rinks and ice fairs, and people fish through holes 6.5–10 ft deep into the flowing water below.

Amazon

- **The Amazon River** flows eastward right across South America, from Peru through Colombia and Brazil to the Atlantic, with tributaries in Ecuador, Venezuela, and Bolivia.

- **At a usually accepted length** of 4,000 mi it is the second longest river in the world. Its source is not far from the Pacific Ocean.

- **It flows into the Atlantic Ocean** near the city of Macapá in Brazil, where the estuary is 200 mi wide.

- **It carries more water** into the sea than any other river in the world, averaging about 7 million cu ft per second.

- **This huge volume of water** is more than all the water carried by the next seven biggest rivers in the world added together.

- **About 20 percent** of all the world's freshwater entering the sea comes from the Amazon, and it is still mixing with salt water about 250 mi from the shore.

- **The area of South America** drained by the Amazon and all its tributaries is 2,700,000 sq mi—about 40 percent of the continent.

- **The source of the Amazon** is Mount Nevado Misimi, high in the Andes Mountains, about 17,000 ft above sea level.

- **Normally between** 1–6 mi wide, during the wet season the Amazon grows to perhaps 30 mi wide and 30 ft deeper than normal.

- **Much of the drainage area** of the Amazon is rainforest, covering 2,000,000 sq mi. Over one-third of all known animal species in the world live there.

◀ *The Amazon River carries more water than any other, and winds thousands of miles through dense rain forest.*

Nile

- **The Nile River flows** northward across the desert in the north-east corner of Africa.

- **At approximately 4,132 mi**, the Nile is usually considered the longest river in the world

- **The Nile and it tributaries** pass through 11 countries: Tanzania, Uganda, Rwanda, Burundi, Democratic Republic of the Congo, Kenya, Ethiopia, Eritrea, South Sudan, Sudan, and Egypt.

- **The Nile basin** drains an area of 1,254,800 mi, which is about 10 percent of the continent of Africa.

- **The White Nile** and the Blue Nile are the two main tributaries, and they join at Khartoum, the capital city of Sudan.

- **The exact source of the White Nile** is still unclear, and so is the exact length of the river. It is near the great lakes in central Africa, in either Burundi or Rwanda.

- **Beginning at Lake Tana** in Ethiopia, the Blue Nile carries most of the water and sediment that fertilizes the desert further downstream.

- **The Nile empties** into the Mediterranean Sea at one of the largest deltas in the world, a silt triangle 150 mi wide and 100 mi long.

- **Ancient Egyptian civilization** thrived because the Nile flooded every year bringing rich, dark, fertile soil, so farmers could grow crops in the desert on either side of the river.

- **In the 1960s** a dam was built across the river at Aswan, giving rise to one of the largest man-made lakes in the world, Lake Nasser.

- **The dam prevented** the annual floods but has allowed irrigation of the riverside farmlands in Egypt to be controlled. However the Nile Delta is under threat from rising sea levels, one of the effects of global warming.

◀ Water from the Nile (far left) is taken by pipes and ditches to irrigate crops and pasture along the banks. Exactly where the irrigation stops, the desert starts.

Mississippi

● **The entire Mississippi River** system has the fourth largest drainage area in the world, covering 40 percent of U.S.A. It has tributaries in Canada and 31 U.S states.

▼ *The city of New Orleans, home to more than one million people, sees huge ships carry cargo between the Mississippi and the Gulf of Mexico.*

- **It begins at Lake Itasca**, Minnesota, and flows southward across the U.S.A. to the Gulf of Mexico near New Orleans.

DID YOU KNOW?
The Mississippi's outlet has one of the biggest river deltas in the world, at more than 4,600 sq mi.

- **The Mississippi is 2,340 mi long**. One drop of water takes about 90 days to get from its source to the sea.

- **The river flows through** or along the borders of Minnesota, Wisconsin, Iowa, Missouri, Kentucky, Tennessee, Arkansas, Mississippi, and Louisiana.

- **Because of the silt** the river has brought down from the Rocky Mountains, the Mississippi River Valley is perhaps the most fertile area of North America.

- **There are 72 dams** along the river's length, which provide electricity, aid navigation, and help to control flooding.

- **The river varies** in width from 20–30 ft where it leaves Lake Itasca, to 41 mi at Lake Winnibigoshish, Minnesota.

- **At St. Louis**, the Mississippi is joined by the Missouri River. Together they comprise the fourth longest river system in the world, at 3,709 mi.

- **Surface speed** of the river increases from one mph near its source to 3 mph at New Orleans.

- **The Mississippi** carries between 247,000 and 706,000 cu ft of water per second to the sea.

Congo

- **Once called the Zaire River**, the Congo drains an area of 1,544,400 sq mi in ten East African countries.

- **It is the ninth longest river** in the world, but at 722 ft in places, it is the deepest.

- **The Congo carries** the second largest amount of water into the sea, an average of 1,448,000 cu ft per second.

- **Only the Amazon rain forest** is bigger than the Congo rain forest, which grows along the banks of the river and its tributaries.

- **The river's drainage area** straddles the Equator so there is always a rainy season somewhere along its length to keep the water flowing.

- **The source of the river**, that is, its longest tributary, is the Chambeshi River in Zambia.

- **The Livingstone Falls** set of waterfalls and rapids in the Democratic Republic of the Congo has more water flowing over it than any other waterfall.

- **This powerful river** provides electricity from about 40 dams. The twin Inga Falls dams, part of the Livingstone Falls near Kinshasa, are the largest.

- **One of the tributaries**, the Lualaba River, flows from Lake Tanganyika and Lake Mweru, almost across the continent near the East African Rift Valley.

- **Most of the river is navigable** between waterfalls, and carries much of the region's produce of copper, palm oil, sugar, cotton, and coffee for trade.

▼ *After the seven cataracts (fast rapids-waterfalls) of the Boyoma Falls, in total covering 62 mi, the Lualaba River joins the Congo. Here a fisherman casts his net into the rapids.*

Yangtze

▲ *The Yangtze River is the world's third longest, flowing east over 3,728 mi from the Tibetan Plateau to the East China Sea.*

- **China's Yangtze River** is known as the Chang Jiang which means "long river."

- **At a generally accepted 3,900 mi** it is the longest river in Asia and the third longest in the world.

- **It begins as meltwater** flowing from glaciers high in the Tanggula Mountains, on the Qinghai-Tibet plateau, and flows eastward across China to Shanghai.

- **One-third of China's population**—nearly half a billion people— live along its valleys and banks.

- **The Yangtze drains** one-fifth of China, an area measuring about 2,000 by 600 mi.

- **It carries an average** one million cu ft of water into the East China Sea, which makes it the largest river by discharge in China and seventh largest in the world.

- **The only way to cross the river** between Yibin and Shanghai was by ferry until 1957 when the first bridge, at Wuhan, was built.

- **The river was dammed** by the Three Gorges Dam project near Yichang, to produce electricity, aid shipping, and control flooding.

- **The dam was started in 1994** and finished in 2009. It raised the level of the river running through the Yangtze Gorges to 574 ft above sea level.

- **Over one million people** had to be rehomed as the water rose in the reservoir and 1,300 historical sites were flooded.

Ganges

- **The Ganges River** begins in the Himalayas and flows for 1,570 mi east and south across India and Bangladesh.

- **The river is fed by meltwaters** from snow on Himalayan mountains such as Nanda, Devi, Trisul, and Kamet.

- **Much of the river's water** comes from the annual summer monsoons when 84 percent of the total yearly rain falls.

- **The Ganges joins** the other major Asian river, the Brahmaputra, at its delta at Dacca in Bangladesh. Together they have formed a complicated and shifting system of channels to the sea.

- **Together with the Meghna River** they carry almost 1,342,000 cu ft of water per second into the Bay of Bengal in the Indian Ocean, making them the third largest river system by discharge in the entire world.

- **The Ganges is sacred** to Hindu people who worship it as the goddess Ganga.

- **In 2007** the Ganges held the unenviable title of fifth most polluted river in the world. A plan to clean it up, the Ganga Action Plan, has had limited effect so far.

- **The river is used** to remove both untreated sewage and dead bodies from the towns and cities along its shores.

- **Despite this**, Hindus believe Ganges river water will purify their bodies and souls, and many bathe in it every day.

- **The Ganges has been used** for irrigation for thousands of years, but as yet, only about 12 percent of its electricity-generating potential has been harnessed.

▼ *Many huge ceremonies and festivals are held along the Ganges. The Kumbh Mela mass Hindu pilgrimage in 2013 attracted over 75 million people.*

Yenisei

- **At 2,166 mi**, The Yenisei River is the fifth longest in the world.

- **It flows from Mongolia** north across Russia to the Yenisei Gulf in the Kara Sea at the edge of the Arctic Ocean.

- **The river rises 11,000 ft** above sea level in the Hangayn Mountains of Mongolia and flows northward.

- **It drains an area** of one million sq mi, which is the seventh largest river drainage area in the world.

- **An average of 706,000 cu ft** of water per second flow from the Yenisei, making it the largest river flowing into the Arctic Ocean.

- **Hydroelectric dams** along its turbulent upper reaches provide much of the power for Russian industry.

- **The river is frozen in the winter**. In spring the upstream water, which is further south and so warmer, thaws out and floods over the still-frozen lower reaches.

- **When it is not frozen**, the river is used to transport wood, grain, and construction materials.

- **The Yenisei has** a serious pollution problem and is discharging pesticides, herbicides, heavy metals, and radioactive waste into the Arctic Ocean.

- **The river was followed** by a team of people along its entire course for the first time in 2001.

- **It is thought** that the original North American people may have come from the Yenisei area across the Bering land bridge about 15,000 years ago.

▼ The upper reaches of the Yenisei pass through Central Asian woodlands that flood in spring.

How lakes form

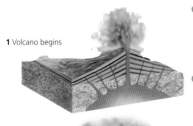

1 Volcano begins

2 Mass eruptions destroy volcano

3 Land collapses as a caldera

4 Caldera fills with water

● **A lake is a large depression** on land that is usually filled with freshwater that doesn't flow, or flows exceedingly slowly.

● **When tectonic activity** uplifts a mountain range, river water can flood the basins left between the peaks. Lake Victoria in Africa is an example of this.

● **Water sometimes collects** in rift zones where two tectonic plates are separating, as with Lake Baikal in Russia.

● **As glaciers flow** toward the sea they scrape depressions in the land which fill with meltwater; the North American Great Lakes formed this way.

● **If a lake has no inlet** or outlet flow, the water will evaporate until the lake becomes too salty for life. One example of this is the Dead Sea.

◀ *A caldera lake forms in the giant hole caused when land collapses into the empty magma chamber of a volcano.*

- **An oxbow lake** forms when a meandering river takes a new course, leaving behind an isolated crescent-shaped lake.

- **The craters or calderas** of inactive volcanoes soon fill with rainwater, as in Crater Lake, Oregon, U.S.A.

- **Soluble rock**, such as limestone, may collapse and leave a sinkhole lake, such as Lake Jackson in Florida, U.S.A.

- **In cold, mountainous places**, ice and snow can sometimes build up and dam a river, which then forms a proglacial lake behind the blockage. Russell Fjord in Alaska is an example of this.

- **There are many artificial lakes** or reservoirs all over the world where rivers have been dammed to produce hydroelectric power.

- **Kariba Lake** on the Zambezi River in central-south Africa is the largest of these dammed artificial lakes. It is 140 mi long and 22 mi wide, with an area of 2,160 sq mi.

▼ *This river in southwest Spain may soon break through the neck of its meander, leaving an isolated oxbow lake.*

Great Lakes

- **The five Great Lakes** are situated in northeast North America, on the border between the U.S.A. and Canada.

- **From west to east** they are Lake Superior, Lake Michigan, Lake Huron, Lake Erie, and Lake Ontario, and they are connected to the sea by the St. Lawrence Seaway.

- **They are the largest group** of freshwater lakes in the world.

- **Together they contain** 5,439 cu mi of water, which is about one-fifth of all the freshwater on the Earth's surface.

▼ The Great Lakes of North America hold 21 percent of the Earth's freshwater.

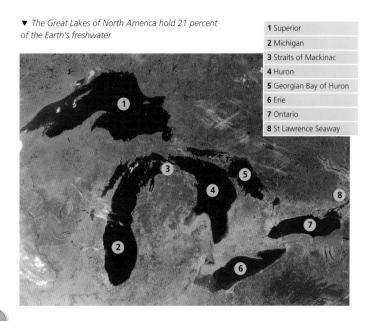

| **1** Superior |
| **2** Michigan |
| **3** Straits of Mackinac |
| **4** Huron |
| **5** Georgian Bay of Huron |
| **6** Erie |
| **7** Ontario |
| **8** St Lawrence Seaway |

▼ *An aerial shot of the Canadian section of Niagara Falls, also known as Horseshoe Falls, with snow still present in early spring. The Niagara River is the natural outlet from Lake Erie to Lake Ontario.*

- **Their total surface area** is 95,000 sq mi, about the same area as the whole of the U.K.

- **If all the water in the lakes** was allowed to flow evenly over North America it would cover the land to a depth of 5 ft.

- **Lake Superior's surface** is 31,660 sq mi, making it the largest lake by area in the world. It is 1,335 ft at its deepest point.

- **There are 35,000 islands** in the Great Lakes and several thousand smaller lakes in the area around them.

- **The lakes and their islands** have approximately 10,500 mi of shoreline, although this is difficult to measure accurately.

- **The Great Lakes formed** only about 10,000 years ago as moving ice glaciers gouged great hollows in the rocks at the end of the last Ice Age.

Lake Baikal

- **Situated in southern Siberia**, Russia, Lake Baikal is the seventh largest lake in the world by surface area, at 12,239 sq mi.

- **It is the world's deepest lake**, at a maximum 5,380 ft. So it holds 5,662 sq mi, which is 20 percent of the world's freshwater that is not permanently frozen.

- **Shaped like a crescent**, Baikal is 30 mi wide and 395 mi long, and it is surrounded on all sides by mountains.

- **The floor of the lake** is covered with sediments up to 4 mi deep.

- **Lake Baikal** is one of the oldest lakes in the world. It was formed 25 million years ago in what was then the deepest continental rift valley in the world.

- **The rift valley** is still getting wider by about 0.8 in every year and there are hot springs and frequent earthquakes in the area.

- **The lake is fed** by about 300 rivers and drains into the Angara River, a tributary of the Yenisei River.

- **Much of Baikal's surface** freezes every year in January and thaws in May or June. The water is very clear and the pollution levels are very low.

- **Baikal is a UNESCO World Heritage Site**. More than 1,000 species of plants and animals living in the lake are not found anywhere else in the world, including the only totally freshwater seal, the Baikal seal.

▶ *Baikal seals, numbering around 90,000, are among the smallest of all seals at just 4 ft long and 140 lb in weight.*

DID YOU KNOW?

Baikal is so deep that its lake bed is 3,887 ft below sea level.

Lake Victoria

- **Lake Victoria** is part of the African Great Lake system, situated across the borders of Uganda, Kenya, and Tanzania.

- **It was named after Queen Victoria** by English explorer John Hanning Speke (1827–1864), who sighted it in 1858 while looking for the fabled source of the White Nile River.

▼ More than 200,000 people fish in Lake Victoria, some as tourists visiting for a few days, others hauling in large catches almost every day for their livelihood.

- **With a surface area** of 26,564 mi, it is the largest lake in Africa and the second largest in the world.

- **It is quite shallow** and contains only about 660 cu mi of water, so it is only the ninth largest lake in the world by volume.

- **Thousands of small streams** drain into the lake from its basin, but 80 percent of its water comes from rainfall.

- **The main river** flowing into the lake is the Kagera River and the only river flowing out of the lake is the Nile River.

- **The drainage area**, or basin, covers 92,000 sq mi across Tanzania, Uganda and Kenya.

- **The shoreline measures** 3,000 mi, which includes the shores of 84 islands.

- **The lake formed** 400,000 years ago and has dried out and refilled at least three times since then.

- **The lake has a serious problem** with water hyacinth, an invasive plant that forms impenetrable floating mats of vegetation.

- **The introduction** of the large fish called the Nile perch has caused extinction of hundreds of species that lived only in the lake.

DID YOU KNOW?
Lake Victoria is by far the shallowest lake for its size, with a greatest depth of just 262 ft.

Lake Tanganyika

- **One of Africa's Great Lakes**, Lake Tanganyika contains 4,544 cu mi of freshwater, making it second only to Lake Baikal in terms of volume.

- **It is up to 4,800 ft deep**, again second only to Lake Baikal in terms of depth.

- **Because of the great depth** the water circulation is poor and the water near the bottom is "fossil" water, having been there mostly undisturbed for thousands of years.

- **The lake is bordered** by Tanzania, the Democratic Republic of the Congo, Burundi, and Zambia.

- **The main outflow** is the Lukunga River, which flows into the Congo River and westward, to enter the Atlantic Ocean at Muanda in the Democratic Republic of the Congo.

- **Lake Tanganyika is crescent-shaped**, 419 mi long from north to south and an average of 30 mi wide.

- **It fills part of the Great Rift Valley** where the African tectonic plate is splitting apart at a rate of about 5 mm each year, to form two new plates, the Somali and Nubian plates.

- **The lake's catchment area** covers 89,000 sq mi and includes hills and plateaus, plains, scattered woods, and dense forests.

- **The first Europeans** to visit the Lake were English explorers Richard Francis Burton and John Hanning Speke in 1858, while searching for the source of the White Nile River.

▶ *A satellite image of Lake Tanganyika. About 100,000 people work in the fishing industry on the lake. About one million people eat these fish, and ten million people depend on the lake and its basin for their livelihoods.*

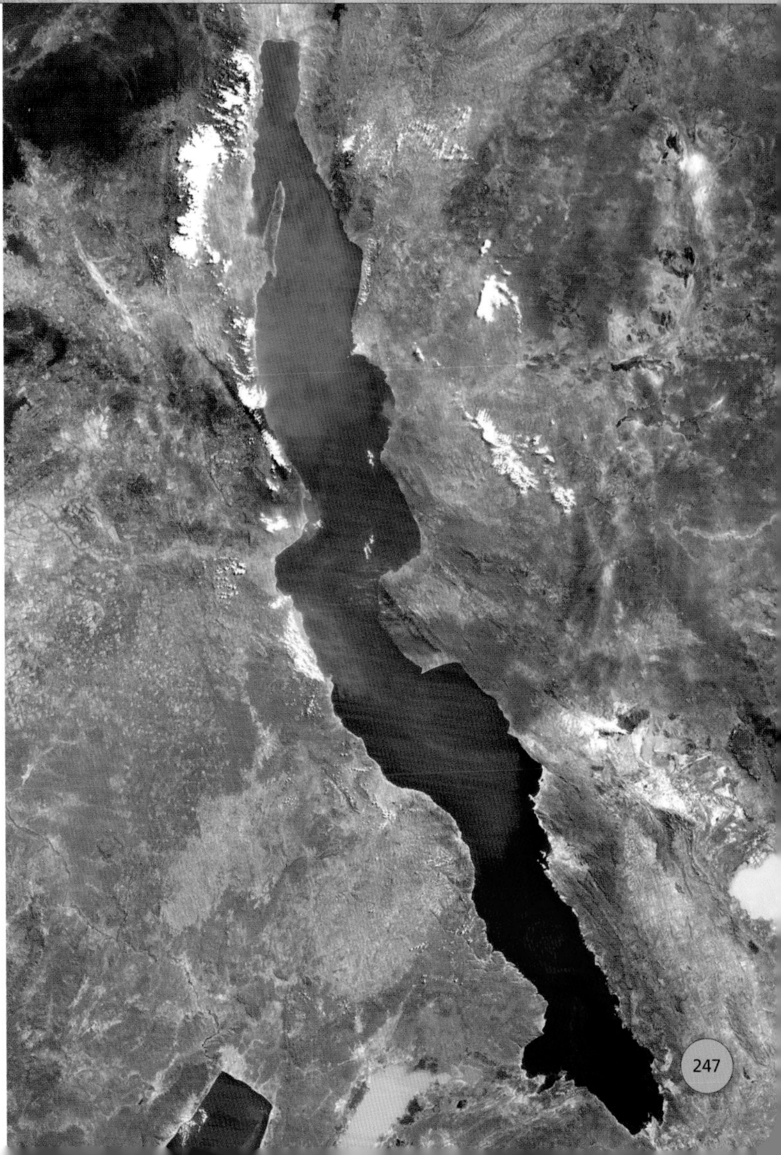

Great Bear Lake

- **Great Bear Lake** lies on the Arctic Circle, in the northern state of Northwest Territories, Canada.

- **The surface area** is 12,000 sq mi and the surface is covered by ice from November to July every year.

- **It is the largest lake** completely in Canada by surface area, not counting the Great Lakes, which are shared with the U.S.A.

- **It is also the third largest lake** in North America by surface area and the eighth largest in the world.

- **The average depth** of the Great Bear Lake is only 230 ft although the deepest point is considerable for a freshwater lake at 1,378 ft.

- **The volume of water** in the lake is 652 cu mi. It takes about 130 years for the water in the lake to move through it, which is called its "residence time."

- **The shoreline measures** 1,700 mi including the many small islands.

- **Only about 500 people** live around Great Bear Lake for the whole year.

- **The catchment area** is 114,700 sq km, being part of the Mackenzie River drainage basin.

- **The name of the lake** comes from the native American people who once lived nearby, called the Chipewyan, which means "grizzly bear people."

- **Settlements on the eastern shore**, called Port Radium, were set up in the 1930s to mine pitchblende, an ore of the radioactive metal uranium, which was used in atomic bombs. They closed after a few years.

▼ Port Radium, now mainly a ghost town, had its own airstrip, shown here, and ferry port during the Great Bear Lake uranium-mining boom times of the 1930s.

DID YOU KNOW?

In 1995 a trout weighing 72 lb was caught in the lake—the largest rod-caught trout in the world.

Lake Vostok

- **There are about 400 lakes** under the ice cap of Antarctica, which no one has ever seen. Lake Vostok is the largest, in fact it is one of the largest lakes in the world.

- **There is 13,000 ft of ice** above the lake's surface, which itself is 1,640 ft below sea level.

- **The lake is about 155 mi long**, up to 30 mi wide, and contains some 1,295 cu mi of water.

- **The average depth** is 1,400 ft and the deepest point is 2,953 ft.

- **Russian scientists** found Lake Vostok by using seismic soundings. Along with British scientists, they confirmed its existence in 1993.

- **All the lakes of Antarctica** may be connected by rivers that flow under the ice.

- **The "residence"** that an average drop of water stays in Lake Vostok is extraordinarily long at 13,300 years.

- **There is a layer of sediment** 230 ft thick on the bottom of the lake. This may hold clues to past climates and traces of living things that existed before the ice formed 400,000 years ago.

- **The ancient**, undisturbed "fossil" water in the lake has probably been cut off from the atmosphere by ice for between 15 and 25 million years, and may contain previously unknown life forms.

- **In 2012** the ice above the lake was drilled through and an ice core 12,362 ft long, the longest ever, was extracted for the study of climate history.

● **The temperature** of the water in the lake is probably about 26°F but it stays liquid because of the huge pressure from the weight of ice above it.

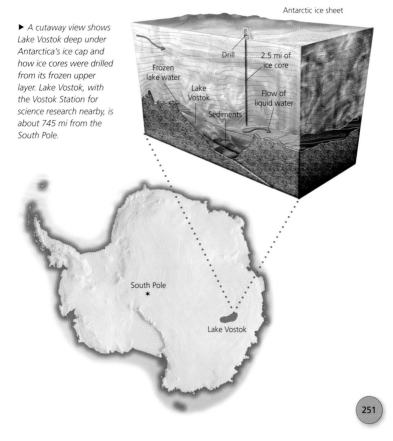

▶ A cutaway view shows Lake Vostok deep under Antarctica's ice cap and how ice cores were drilled from its frozen upper layer. Lake Vostok, with the Vostok Station for science research nearby, is about 745 mi from the South Pole.

Antarctic ice sheet

Drill

2.5 mi of ice core

Frozen lake water

Lake Vostok

Flow of liquid water

Sediments

South Pole
*

Lake Vostok

Atmosphere and weather

Atmosphere

- **The atmosphere is a blanket** of gases that extends to 6,200 mi above the Earth. It can be divided into five layers: troposphere (the lowest layer), stratosphere, mesosphere, thermosphere, and exosphere.

- **The atmosphere is**: 78 percent nitrogen, 21 percent oxygen, one percent argon and carbon dioxide, with tiny traces of neon, krypton, xenon, helium, nitrous oxide, methane, and carbon monoxide. It also contains a variable amount of water vapor.

- **It was first created** by the fumes pouring out from volcanoes that covered the early Earth 4,000 million years ago. But the atmosphere changed as rocks and seawater absorbed carbon dioxide, and then single-celled bacteria and algae in the sea built up oxygen levels over millions of years.

- **The troposphere** is just 0–12 mi thick and yet it contains 75 percent of the weight of gases in the atmosphere.

- **Temperatures in the troposphere** drop with height from an average of 64°F to about −76°F at the top, or tropopause.

- **The stratosphere** contains little water vapor. Unlike the troposphere, which is heated from below, the stratosphere is heated from above as the ozone molecules in it are heated by ultraviolet (UV) light from the Sun. Temperatures rise with height from −76°F at the bottom to 50°F at the top, about 30 mi up.

- **The stratosphere** is usually clear and calm, which is why passenger aircraft fly in this layer.

▶ *The atmosphere is a sea of colorless, tasteless, odorless gases, mixed with moisture and fine dust particles. It is about 620 mi deep but has no distinct edge, simply fading away into space. As you move up, each layer contains less and less gas. The topmost layers are very rarefied, which means that gas is sparse.*

● **The mesosphere** contains few gases but it is thick enough to slow down meteorites. They burn up as they hurtle into it, leaving fiery trails in the night sky. Temperatures drop from 50°F to −184°F about 50 mi up.

● **In the thermosphere** temperatures are high, but there is so little gas that there is little heat. Temperatures rise from −184°F to 3,600°F about 370 mi up.

● **The exosphere** is the highest level of the atmosphere where it fades into the nothingness of space.

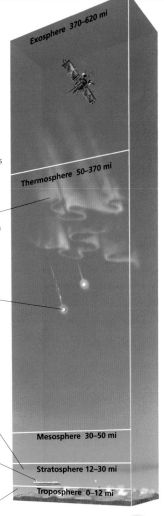

Exosphere 370–620 mi

Thermosphere 50–370 mi

Shimmering curtains of light called auroras appear above the poles. They are caused by the impact of particles from the Sun on the gases in the upper atmosphere

The atmosphere protects us from meteorites and radiation from space

The stratosphere contains the ozone layer, which protects us from the Sun's UV rays

Mesosphere 30–50 mi

Aircraft climb to the stratosphere to find calm air

Stratosphere 12–30 mi

Troposphere 0–12 mi

The troposphere is the layer closest to Earth's surface

255

Lights in the sky

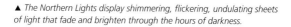

▲ The Northern Lights display shimmering, flickering, undulating sheets of light that fade and brighten through the hours of darkness.

- **The spectacular light show** in the northern night sky is called the *Aurora Borealis* or Northern Lights. In the Southern Hemisphere a similar display of lights is called *Aurora Australis* or Southern Lights.

- **These curtain-like lights,** usually green, sometimes tinged with red, occur in the Arctic and Antarctic near the magnetic North and South Poles.

- **The lights are caused** by electrically charged atoms and molecules in the high atmosphere colliding with similar particles in the solar wind that pours out from the Sun.

- **The solar wind sub-atomic particles** are attracted to the Poles by the Earth's magnetic field.

- **When they collide** with the charged atoms or molecules, energy in the form of photons—packets of light energy—is released.

- **Oxygen is one of the molecules** involved and it gives off green or orange light. Nitrogen is another and it gives off blue or red light.

- **Auroras are more dramatic** during the equinox months in spring and autumn, when the periods of daylight are the same length as the periods of night darkness.

- **When the Sun** is in the sunspot phase of its 11-year cycle, auroras are even more dramatic, and seen over a larger area.

- **One of the most spectacular auroras** ever recorded was in September of 1859 when people in Boston, U.S.A., reported they could read a newspaper at one o'clock in the morning by its light.

- **This event** also disrupted the thousands of miles of newly laid telegraph cables.

Sunshine

- **Half of the Earth** is exposed to the Sun at any time. Radiation from the Sun is the Earth's main source of energy, providing huge amounts of heat and light.

- **"Solar"** means anything to do with the Sun.

- **About 41 percent** of solar radiation is light; 51 percent is long-wave radiation that our eyes cannot see, such as infrared (IR) light, that is, heat. The other 8 percent is short-wave radiation, such as ultraviolet (UV) rays.

- **Only 47 percent** of the solar radiation that strikes the Earth actually reaches the ground. The rest is soaked up or reflected by the atmosphere.

▶ Snow reflects away a lot of the Sun's heat, so the ground beneath stays cold.

- **Solar radiation** reaching the ground is called insolation.

- **The air is not warmed** much by the Sun directly, but by heat reflected from the ground.

- **The amount of heat** reaching the ground depends on the angle of the Sun's rays. The lower the Sun is in the sky, the more its rays are spread, and so give off less heat.

- **Insolation is at a peak** in the tropics and during the summer. It is lowest near the Poles and in winter.

- **Some surfaces** reflect the Sun's heat and warm the air better than others. The percentage they reflect is called the albedo.

- **Snow and ice** have an albedo of 85–95 percent and so they stay frozen even as they warm the air.

- **Forests have an albedo** of 12 percent, so they soak up a lot of the Sun's heat.

Air moisture

- **Up to 6 mi above the ground**, the air is always moist because it contains an invisible gas called water vapor.

- **On average**, air is about one percent water vapor, but the amount varies considerably.

- **Water vapor enters** the air when it evaporates from oceans, rivers, and lakes.

- **It is also given off by plants**, especially leaves and grass, and in air breathed out by animals.

- **Water vapor leaves** the air when it cools and condenses (turns to drops of water) to form clouds. Most clouds eventually turn to rain, and the water falls back to the ground. This is called precipitation.

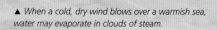

▲ *When a cold, dry wind blows over a warmish sea, water may evaporate in clouds of steam.*

- **Like a sponge**, the air soaks up evaporating water until it is saturated (full). It can only take in more water if it warms up and expands.

- **If saturated air cools**, it contracts and squeezes out the water vapor, forcing it to condense into drops of water. The point at which this happens is called the dew point.

- **Humidity** is the amount of water in the air.

- **Absolute humidity** is the weight of water in grams in a particular volume of air.

- **Relative humidity**, which is written as a percentage, is the amount of water in the air compared to the amount of water the air could hold when saturated.

DID YOU KNOW?

There is enough water vapor in the atmosphere to flood the entire globe to a depth of 8 ft.

261

Clouds

- **Clouds are dense masses** of water drops and ice crystals that are so tiny, they float high in the air.

- **Cumulus clouds** are fluffy and white. They pile up as warm air rises and cools to the point where water vapor condenses.

- **Strong updraughts** create huge cumulonimbus, or thunderclouds.

- **Stratus clouds** are vast, shapeless clouds that form when a layer of air cools to the point where moisture condenses. They often bring long periods of light rain.

- **Cirrus clouds** are wispy, and form so high up, they are made entirely of ice. Strong winds high up blow them into "mares' tails."

- **Low clouds** lie below 6,500 ft altitude. They include stratus and stratocumulus clouds (the spread tops of cumulus clouds).

- **Nimbus are low**, dark clouds that bring rain or other precipitation.

- **Middle clouds** often have the prefix "alto" and lie from between 6,500–19,700 ft up. They include rolls of altocumulus cloud, and thin sheets called altostratus.

- **High-level clouds** are ice clouds up to 36,000 ft up. They include cirrus, cirrostratus, and cirrocumulus.

- **Contrails**, short for condensation trails, are long, thin trails of ice crystals or water droplets made by jet aircraft.

▶ *High level clouds like this are formed more of ice than water. Strong winds high-up create distinctive ripples.*

Fog and mist

- **Like clouds**, mist is made up of billions of tiny water droplets floating in the air.

- **Fog forms** close to the surface of the Earth, over bodies of water or moist ground.

- **Mist forms** when the air cools to the point where the water vapor it contains condenses to water.

- **Meteorologists define fog** as a mist that reduces visibility to less than half a mile.

▶ Mist often forms when cold air sinks into valleys on chilly nights.

- **There are four main kinds** of fog: radiation, advection, frontal, and upslope.

- **Radiation fog** forms on cold, clear, calm nights. The ground loses the heat that it absorbed during the day, and cools the air above.

- **Advection fog** forms when warm, moist air flows over a cold surface. This cools the air so much that the moisture it contains condenses.

- **Sea fog** is advection fog that forms as warm air flows out over cool coastal waters and lakes.

- **Frontal fog** forms along weather fronts.

- **Upslope fog** forms when warm, moist air rises up mountains and then cools.

DID YOU KNOW?
Smog is a thick fog that forms in air polluted with tiny particles.

Rain

- **Rain falls from clouds** filled with large water drops and ice crystals.

- **The technical name** for rain is precipitation, which also includes snow, sleet, hail, frost, and dew.

- **Drizzle has drops** 0.2–0.5 mm in diameter and it falls from nimbostratus clouds. Raindrops from nimbostratus are 1–2 mm in diameter. Drops from cumulonimbus (thunderclouds) can be 5 mm in diameter.

- **Rain starts** when water drops or ice crystals inside clouds grow too large for the air to support them.

- **Water drops** in clouds grow when moist air is swept upward and cools, causing lots of drops to condense and merge. This happens when pockets of warm, rising air form thunderclouds—at weather fronts or when air is forced up over hills.

- **In the tropics**, raindrops grow in clouds by colliding with each other. In cool places, they also grow on ice crystals in the clouds.

- **The world's rainiest place** is Mount Wai-'ale-'ale in Hawaii, where it rains 350 days a year.

- **Rainfall estimates** give the wettest place as Lloro in Colombia, which gets more than 512 in (43 ft) of rain every year.

- **La Réunion island** in the Indian Ocean received 74 in of rain in one day in 1952.

- **Guadeloupe** in the West Indies received 1.5 in of rain in one minute in 1970.

◄ Tall, dark
thunderclouds
unleash rain in
short, heavy
showers.

Snow and hail

- **Snow is a type of precipitation** made of ice crystals. It falls in cold weather when the air is too cold to melt the ice to rain.

- **Outside of the tropics,** most rain starts to fall as snow but melts on the way down.

- **More snow falls** in the northern U.S.A. than at the North Pole. It is usually too cold to snow at the North Pole.

- **The heaviest snow** falls when the air temperature is hovering around freezing.

- **Snow can be hard to forecast** because a rise in temperature of just 1°F can turn snow into rain.

- **All snowflakes have six sides**. They usually consist of crystals that are flat plates, but occasionally needles and columns are found.

- **Wilson Bentley** (1865–1931) was an American farmer who photographed thousands of snowflakes through microscopes. He never found two identical flakes.

◀ The mist above the hills here is not clouds, but tiny ice crystals blown off the snow.

- **Hail is in effect frozen raindrops**, usually ball-shaped and larger than 4–5 mm.

- **The balls are called hailstones** and when many fall in a short time this is a hailstorm.

- **Most hailstones have onionlike layers**. These are due to freezing from raindrops, then being swept up and down inside a cloud, as more water droplets and water vapor condense and freeze onto the ball each time.

- **Hailstones larger** than about 0.8 in can damage cars, roofs, windows, and other structures.

- **The heaviest hailstone** scientifically recorded weighed more than 2 lb and fell in Bangladesh. The largest was 8 in across, and fell in South Dakota, U.S.A.

DID YOU KNOW?
The city that recieves the most amount of snow is Aomori in Japan, with more than 26 ft yearly.

269

Ice and cold

- **Winter is cold** because the days are short. In winter, the Sun moves across the sky at a low angle, so its warmth is spread out.

- **The coldest places** in the world are around the North and South poles. Here, the Sun shines at a low angle even in summer, and in winter it does not rise for days on end.

- **The average temperature** at Polus Nedostupnosti (Pole of Cold) in Antarctica is –72°F.

- **The coldest temperature** ever recorded was –128.5°F at Vostok in Antarctica on July 21, 1983.

- **The interiors of the continents** can get very cold in winter because land loses heat rapidly and there are no warmer oceans nearby.

- **When air cools below freezing point** (32°F), the water vapor in it may freeze without turning first to dew (water drops that form on cool surfaces). It covers the ground with crystals of ice or frost.

- **Fern frost** is feathery tails of ice that form on cold glass as dew drops freeze bit by bit.

- **Hoar frost** forms as spiky needles when damp air blows over cold surfaces and freezes onto them.

- **Rime is a thick coating** of ice that forms when drops of water in clouds and fogs remain as liquid well below freezing point. The drops freeze instantly when they touch a surface.

- **Black ice forms** when rain falls on freezing road surfaces.

▼ *A sudden plunge in temperature turns a waterfall into a curtain of icicles.*

Air pressure

- **Air is light**, but there is so much that it can exert huge pressure at ground level. Air or atmospheric pressure is the constant bombardment of billions of air molecules as they move around.

- **Air pushes in all directions** at ground level with a force of over 2.2 lbs per square centimeter—that is the equivalent of an elephant standing on a coffee table.

▼ Familiar from weather charts and forecast maps, these white lines are isobars. They represent places of equal air pressure (iso = same or equal, bar = pressure). So the 980 line shows all places with an air pressure of 980 millibars. The closer the isobars, the stronger the winds in that area. "L" signifies the middle of this low pressure system—a hurricane in the Gulf of Mexico.

1012
1004
996
988
980
L

- **A device called a barometer** measures air pressure in units called millibars (mb).

- **Air pressure varies constantly** in different places and at different times. This is because the Sun's heat varies too, making air expand to get thinner and lighter, or contract to get thicker and heavier.

- **Normal air pressure** at sea level is 1,013 mb, but it can vary between 800 mb and 1,050 mb.

- **Air pressure is shown** on weather maps by lines called isobars, which link locations that have equal pressure.

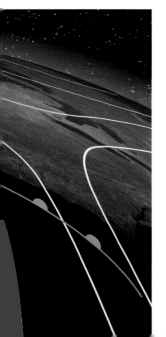

- **High-pressure zones** are called anticyclones. Low-pressure zones are called cyclones or depressions.

- **Barometers help to forecast** weather because changes in air pressure are linked to changes in weather.

- **A sudden, sharp fall** in air pressure warns that stormy weather is on its way because depressions are linked to storms.

- **Steady high pressure** indicates clearer, calm weather.

Weather fronts

- **A weather front** is where a large mass of warm air meets a large mass of cold air.

- **At a warm front**, the mass of warm air is moving faster than the cold air. The warm air slowly rises over the cold air, sloping gently up to one mi over a distance of 186 mi.

- **At a cold front**, the mass of cold air is moving faster. It undercuts the warm air, forcing it to rise sharply and creating a steeply sloping front. The front climbs to one mi over a distance of about 62 mi.

- **In the mid-latitudes**, fronts are linked to vast spiraling weather systems called depressions, or lows. These are centered on a region of low pressure where warm, moist air rises. Winds spiral into the low—anticlockwise in the Northern Hemisphere, clockwise in the Southern.

- **Lows start along** the polar front, which stretches around the world. Here, cold air from the poles meets warm, moist air moving up from the subtropics.

- **Lows develop as a kink** in the polar front. They then grow bigger as strong winds in the upper air drag them eastward, bringing rain, snow, and blustery winds. A wedge of warm air intrudes into the heart of the low, and the worst weather occurs along the edges of the wedge. One edge is a warm front, the other is a cold front.

- **The warm front arrives** first, shown by feathery cirrus clouds of ice high in the sky. As the front moves over, the sky fills with slate-gray nimbostratus clouds that bring steady rain. As the warm front passes away, the weather becomes milder and skies may briefly clear.

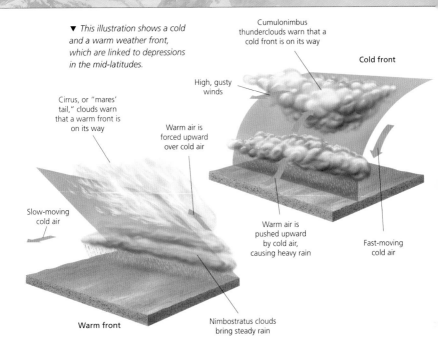

▼ *This illustration shows a cold and a warm weather front, which are linked to depressions in the mid-latitudes.*

Cumulonimbus thunderclouds warn that a cold front is on its way

Cold front

High, gusty winds

Cirrus, or "mares' tail," clouds warn that a warm front is on its way

Warm air is forced upward over cold air

Slow-moving cold air

Warm air is pushed upward by cold air, causing heavy rain

Fast-moving cold air

Warm front

Nimbostratus clouds bring steady rain

- **After a few hours,** a build-up of thunderclouds and gusty winds warn that the cold front is on its way. When it arrives, the clouds unleash short, heavy showers, and sometimes thunderstorms or even tornadoes.

- **After the cold front passes,** the air grows colder and the sky clears, leaving just a few fluffy cumulus clouds.

- **Meteorologists think** that depressions are linked to strong winds, called jet streams, which circle the Earth high above the polar front. The depression may begin with Rossby waves, which are giant kinks in the jet stream up to 1,245 mi long.

275

Wind

- **Wind is moving air.** Strong winds are fast-moving air and gentle breezes are air that moves slowly.

- **Air moves** because the Sun warms some places more than others, creating differences in air pressure.

- **Warmth makes air expand** and rise, lowering air pressure. Cold makes air heavier, raising pressure.

- **As air warms and rises,** cooler air flows along to take its place, creating winds.

- **Winds blow** from areas of high pressure, called highs, to areas of low pressure, called lows.

- **The sharper the pressure difference,** or gradient, the stronger the winds blow.

- **In the Northern Hemisphere,** winds spiral clockwise out of highs, and anticlockwise into lows. In the Southern Hemisphere, the reverse occurs.

- **A prevailing wind** is one that blows frequently from the same direction. Winds are named by the direction they blow from. A westerly wind blows from the west.

- **In the tropics** the prevailing winds are warm and dry. They blow from the northeast and the southeast toward the Equator.

- **In the mid-latitudes** the prevailing winds are warm, moist westerlies.

DID YOU KNOW?
The windiest place is George V in Antarctica, where 200 mph winds are common.

▼ The most common or "prevailing" wind direction is different in the tropics, in the mid-latitudes and in the polar regions.

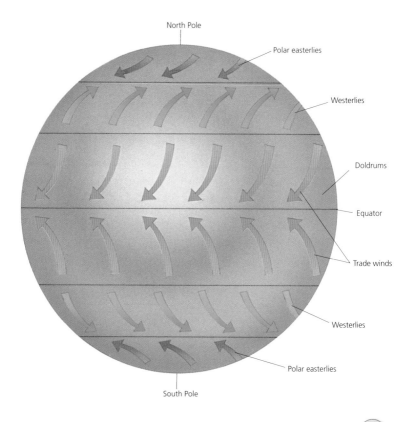

North Pole

Polar easterlies

Westerlies

Doldrums

Equator

Trade winds

Westerlies

Polar easterlies

South Pole

Thunderstorms

- **Thunderstorms begin** when strong updraughts build up towering cumulonimbus clouds.

- **Water drops** and ice crystals in thunderclouds are buffeted together. They become charged with static electricity.

- **Negative charges** in a cloud sink and positive charges rise. Lightning is a surge of negative charge traveling toward the positive.

- **Sheet lightning** is a flash within a cloud. Fork lightning is a flash that travels from a cloud to the ground or to another cloud.

- **Fork lightning** begins with a fast, dim flash from a cloud to the ground, called the leader stroke. It prepares the air for a bigger, slower stroke a split second later.

- **Fork lightning travels** at up to 62,000 mi per second down a narrow path up to 9 mi long. Sheet lightning can be 87 mi long.

- **Thunder is the sound** of the shock wave as air expands when heated instantly to 45,000°F by the lightning.

- **Sound travels slower than light**, so we hear thunder three seconds later for every 0.6 mi between us and the storm.

- **At any moment** there are 2,000 thunderstorms around the world. Every second, 100 lightning bolts hit the ground.

- **A lightning flash** is brighter than ten million 20-watt low-energy light bulbs. For a split second it has more power than all the power stations in the U.S.A. put together.

▼ *The tremendous turbulence in a thundercloud builds up an electrical charge that is released in flashes of lightning.*

Dust and sand storms

- **Dust and sand storms** occur in dry or arid places when the wind picks up loose particles, carries them, and drops them in another place.

- **When the wind blows** larger loose particles of sand, they vibrate on the ground at first and then start to bounce along. This is called saltating.

▼ A looming wall of sand and dust blows toward the pyramids at Giza in Egypt. In minutes visibility will reduce to a few feet.

- **As they bounce** they break up into even smaller particles, or dust, and these can be light enough to get picked up and carried by the wind as dense clouds that lead to poor visibility.

- **The smaller the particles**, the higher they can be lifted in the air, up to 4 mi, and can be carried thousands of miles. The Amazon rain forest receives some of its nutrients from Saharan dust blown across the Atlantic.

- **There are ten times** as many Saharan dust storms now as there were 50 years ago, probably due to poor farming practices such as leaving ground without vegetation cover.

- **Dust storms can cause** diseases such as silicosis and lung cancer, and can affect the eyes, leading to blindness. They can also carry virus spores between continents.

- **In the U.S.**, the Dust Bowl period of the 1930s refers to a series of severe dust storms. Much of the fertile topsoil of the North American prairies was blown away after a severe drought.

- **In July 2014** Sydney, Australia, all but disappeared under a massive cloud of red dust blown eastward from the outback.

- **Parts of the Great Wall of China** have been eroded by sand storms caused by the desertification of the northwest of the country.

- **One of the world's worst sand storms** hit Iraq in 2009. It lasted one week, closed businesses, shops, roads, and the airport, and kept people indoors. The number of people attending hospital for breathing problems was eight times higher than usual.

Blizzards and whiteouts

- **A blizzard is a snow storm** in which snow falls heavily for several hours, with sustained winds of 30 mph or more.

- **The winds whip** around the falling snow and even pick up already-fallen snow from the ground and blow that along.

- **Windblown snow** collects in piles called snowdrifts than can be dozens of feet deep, even burying whole houses.

- **A whiteout** is when falling snow is so thick and/or windblown that visibility is greatly reduced to 3–7 ft.

- **With no visible reference points,** such as the horizon, houses, trees, roads, even the sky and ground, people can become extremely disorientated.

- **Light, fluffy, powdery snow** melts to produce 1/15th–1/10th the equivalent amount of rain, so on average, 5 in of this snow equals 0.3 in of rain. Slushy, dense snow is the equivalent to one-fifth the amount of rain.

- **In February 1972,** almost 26 ft of snow fell on parts of Iran over the period of a week. More than 4,000 people died.

- **In 2014,** a roof collapsed under the weight of snow at a university building in South Korea, killing ten people and injuring another 100.

- **In February 1959**, the Mount Shasta Ski Bowl in California, U.S.A., had 16 ft of snow in just six days.

- **In March 1911**, Tamarac, California, was buried in 36 ft of snow.

- **The snowline** is the lowest level on a mountain where snow remains throughout summer. It is 16,400 ft in the tropics, 8,800 ft in the Alps, 2,000 ft in Greenland, and at sea level at the poles.

◀ Snow tends to blow up and over objects on the windward side, which is to the left in this photograph. It leaves those areas partially uncovered, and then swirls and collects in deep piles on the leeward or downwind side, to the right.

Tornadoes

- **Tornadoes, or twisters**, are tall funnels of violently spiraling winds beneath thunderclouds.

- **A tornado can roar** past in just a few minutes, but it can cause severe damage.

- **Wind speeds** inside tornadoes are difficult to measure, but they are believed to be over 250 mph.

- **Tornadoes form** beneath huge thunderclouds, called supercells, which develop along cold fronts.

- **England has more tornadoes** per square mile than any other country, but they are usually very small and mild.

◀ A waterspout is a weaker tornado that develops over water. Waterspouts have been known to transport objects such as fish and frogs.

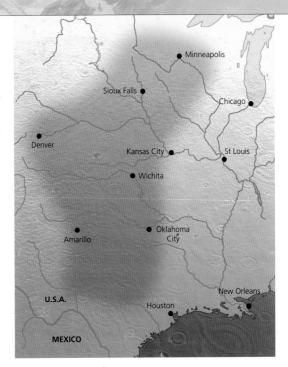

► Tornado Alley (shaded area) is the broad region of central U.S.A. that is hit every summer by numerous tornadoes.

- **In the U.S.A.**, Tornado Alley is centered on the states of Kansas and Oklahoma, and has about 1,000 tornadoes a year. Some of them are incredibly powerful.

- **A tornado may be rated** on the Fujita scale, from F0 (F zero, gale tornado) to F6 (inconceivable tornado).

- **An F5 tornado** (incredible tornado) with wind speeds of 260–315 mph can lift up a house and carry a bus hundreds of feet.

- **In 1990**, a Kansas tornado lifted an 88-carriage train from the track and dropped it in piles, four cars high.

Hurricanes

- **Hurricanes are powerful**, whirling tropical storms. They are also called cyclones (Indian Ocean and southwest Pacific) or typhoons (northwest Pacific). In the Atlantic and Northeast Pacific, they are simply known as hurricanes.

- **In late summer** Atlantic hurricanes develop as clusters of thunderstorms build up over warm seas, at least 80°F.

- **As hurricanes grow**, they tighten into a spiral with a calm ring of low pressure, called the "eye," at the center.

- **Hurricanes move westward** in the Atlantic at about 12 mph. They strike east coasts of the Caribbean and North America, bringing torrential rain and winds gusting up to 220 mph.

- **A hurricane is a storm** with winds exceeding 75 mph.

- **Each hurricane is given a name**. Every year these run in alphabetical order, from a list of six sets of names issued by the World Meteorological Organization.

▶ Satellite technology allows hurricanes to be tracked minute by minute. This image shows Hurricane Ivan of 2004.

DID YOU KNOW?
On average hurricanes last for 3–14 days. They lose energy as they move toward the poles into cooler air.

April/September 2000

September 13, 2005

- **So the first storm of the year** has a name beginning with "A:" Alberto in 2012, Andrea in 2013, then Arthur, Ana, Alex, and Arlene in 2017. The next storm of the season starts with 'B' and so on.

- **The names of hurricanes** that cause huge damage and loss of life, such as Andrew in 1992 and Katrina in 2005, are "retired" and replaced by new ones.

- **The most fatal cyclone** ever struck Bangladesh in 1970. The flood from the storm surge killed 266,000 people.

- **The worst damage** from a hurricane is often done not by the wind, but by the storm surge.

- **Hurricane Katrina** hit the southern coast of the U.S.A., including the city of New Orleans, between August 23 and 31, 2005. It is one of only four Category 5 hurricanes to hit the U.S.A. Category 5 hurricanes have winds higher than 155 mph, but Katrina had winds of up to 175 mph.

▶ *The city of New Orleans, U.S.A., before and after Hurricane Katrina hit. On the lower image, flooding can be seen as light gray (shallower) to dark blue (deeper) areas inland. Lakefront Airport and Lake Pontchartrain are at the top.*

287

Seasons

● **Seasons are periods** into which the year is divided according to annual shifts in weather patterns.

● **Outside the tropics** there are four seasons each year: spring, summer, fall, and winter. Each lasts about three months.

● **The changes in the seasons** occur because the tilt of the Earth's axis, 23.5°, stays the same as the planet circles the Sun.

● **When the Northern Hemisphere** tilts toward the Sun, it is summer in the north of the world and winter in the south.

Spring in Northern Hemisphere

▶ As the Earth orbits the Sun it is always tilted the same way. This means that each pole leans toward and then away from the Sun over the year, giving us seasons.

Summer in Northern Hemisphere

- **When the Earth moves** another quarter of its orbit, the Northern and Southern Hemispheres are equal and it is fall in the north and spring in the south.

- **After another quarter-orbit**, the Southern Hemisphere is tilted toward the Sun, so it is winter in the north of the world and summer in the south.

- **As the Earth moves** to its three-quarters orbit around the Sun, the north begins to tilt toward the Sun again. This brings warmer spring weather to the north and autumn to the south.

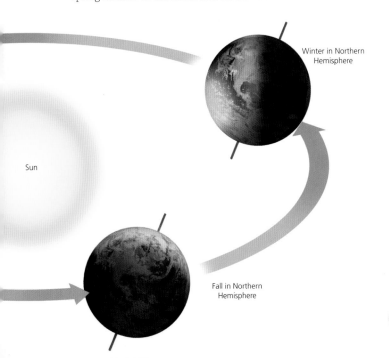

Winter in Northern Hemisphere

Sun

Fall in Northern Hemisphere

Climate zones

- **Weather is atmospheric events** over days and weeks, while climate is longer-term conditions over years, decades, and centuries.

- **Climates are warmer** near the Equator, where the Sun climbs high in the sky. Its rays pass at a steep angle through the atmosphere, and its heat covers a small area on the ground.

- **Tropical climates** occur in the Tropics, the zones either side of the Equator. Temperatures average 80°F.

- **The climate is cool** near the Poles, where the Sun never climbs high in the sky. Its rays slant through the atmosphere and spread over a wide ground area. Temperatures average −22°F.

- **Temperate climates** are mild, and occur in the temperate zones between the tropics and the polar regions. Summer temperatures may average 73°F, and in winter, 53°F.

- **An oceanic climate** is wetter near the oceans, with cooler summers and warmer winters.

- **A Mediterranean climate** is temperate with warm summers and mild winters. It is typical of the Mediterranean, South Africa and South Australia.

- **A monsoon climate** has one very wet and one very dry season—typical of India and Southeast Asia.

- **A continental climate** is dry in the center of continents, with hot summers and cold winters.

- **Mountain climates** get colder and windier with height.

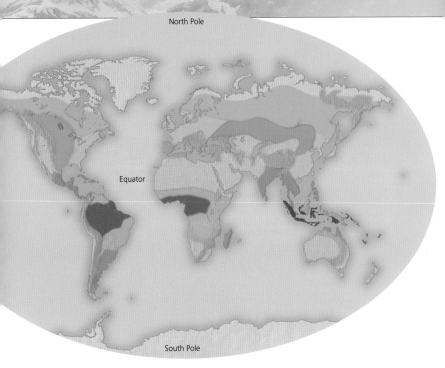

North Pole

Equator

South Pole

Tropical

Tropical rain

Desert

Polar

Mountainous

Dry temperate

Wet temperate

Cold temperate

Temperate grassland

▲ The colored rings (left) match the areas on the map. In general, the warmer climates are found close to the Equator.

291

Weather forecasting

- **The scientific study** of the atmosphere, including weather and climate, is called meteorology.

- **Weather forecasting** relies partly on powerful computers, which analyze the Earth's atmosphere.

- **One kind of weather prediction** divides the air into "parcels." These are stacked in columns above grid points spread throughout the world.

- **There are over one million grid points**, each with a stack of at least 30 parcels above it.

- **Weather stations** take millions of measurements of weather conditions at regular intervals each day.

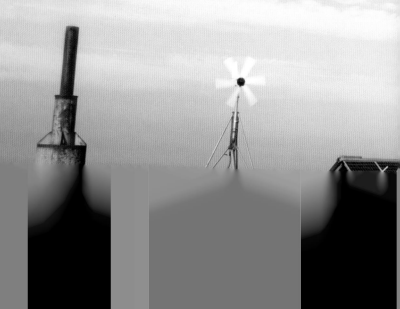

- **Every few hours**, more than 15,000 land-based weather stations record conditions on the ground.

- **Also every few hours**, balloons carry instruments up into the atmosphere to record conditions high up.

- **Meteosats**, or meteorological satellites, give an overview of developing weather patterns.

- **Infrared satellite images** show temperatures on the Earth's surface, on land and water, and in the atmosphere.

- **Supercomputers allow** the weather to be predicted accurately three days in advance, and for up to 14 days in advance with some confidence.

▼ Weather forecasting depends on continual observations from thousands of weather stations around the world, such as this one in the Arctic.

293

Exploiting Earth

Earth's riches

- **Human lives depend** on all sorts of substances found in or on the Earth. Most of these will run out eventually and new sources or alternatives must be found.

- **Fuel is one of the basic requirements** for modern life and comforts. Currently the main fuels are gas, oil, coal, and radioactive elements such as uranium.

- **Precious metals** from the Earth such as gold, silver, and platinum, and gemstones such as diamonds and emeralds, are valued mainly for their rarity and beauty.

- **More common metals** such as iron and copper are used for engineering, technology and building, and constructing the machines we use every day.

- **Rarer metals**, such as scandium and gadolinium, are essential for devices such as cell phones, computers, batteries, and spacecraft.

- **Almost everything** we use contains plastics that are mainly made from petrochemicals (crude oil). New plastics made from plant material are being developed.

- **Up to 50 percent** of the world's crops are grown using fertilizers. Synthetic fertilizers are produced in factories from minerals or natural gas found in rocks.

- **First used 6,000 years ago**, salt is used to season or preserve food, and in the chemicals industry. It is made by evaporating seawater, or mined from rock deposits.

▶ *Extracting raw materials from the Earth involves the world's biggest machines, such as this offshore oil platform.*

- **Constructing buildings** and roads uses substances from the Earth such as sand and gravel, and cement, which is limestone heated to very high temperatures.

- **There is now more copper** in use than there is left in the Earth to mine. As huge amounts of energy are needed to extract aluminum from its ore rocks, metal recycling is now big business.

Surveying and prospecting

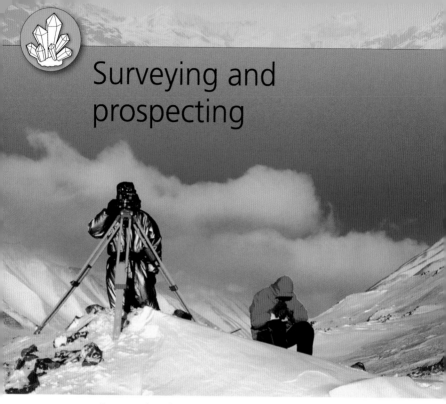

▲ Surveyors using a theodolite to measure horizontal and vertical angles near the Arctic Ocean. As well as aerial and satellite information, traditional instruments are still used.

● **Trying to find the Earth's riches** involves first physically searching the rocks for telltale signs. This stage is called prospecting or fossicking.

● **A geological surveyor** can estimate the presence of underground resources by looking at landforms and surface rock types, their strata (layers), and outcrops.

- **A geophysical survey** is carried out using techniques such as seismology, magnetometry (measuring the Earth's magnetic field), gravimetry (measuring gravity), satellite and aerial photographic images, and radar.

- **Gravity and magnetism** are affected by the type of rock and how it was formed. They can be accurately measured using precision instruments.

- **Seismology involves** analyzing the way sound waves, produced by explosions or machines, travel through rocks to estimate their density and makeup.

- **Still one of the most important** prospecting tools, a metal detector senses changes in the magnetic field. Such devices were first used industrially in the 1960s.

- **If the site looks promising**, a test well may be drilled or a test shaft dug, to verify the results of the initial prospecting.

- **Once a potential site is found**, a survey is done to decide what is the best method of extraction, the machines and construction required, who owns the site, and any safety issues.

- **Before recovery** of the resource can begin, the extraction company must compare its current and possible future value with potential costs.

DID YOU KNOW?

Satellite photography can pinpoint mineral-rich rocks, while devices called spectroscopes can even identify the minerals.

Metals and minerals

- **There are about 91** known metal elements. They are mostly malleable (can be shaped by hammering), fusible (can be fused or melted), or ductile (able to be drawn out into a thin wire).

- **Metals are found** combined with other chemicals in rocks called ores in the Earth's crust. Ores are mined and the pure metals extracted usually by heating, called smelting.

- **Minerals are chemical compounds** that make up the rocks of the crust. About 90 percent of them are silicate minerals—made up primarily of silicon and oxygen.

- **Mineral hardness** is measured on the Mohs Scale. The softest, talc, scores 1, quartz (sand) scores 7 and diamond, the hardest, scores 10.

▼ Quarrying and mining are huge businesses. Here an iron ore rock called haematite is being quarried using excavators, diggers, massive trucks, and railroad wagons.

- **The first metals to be mined** were copper and tin about 5,000 years ago. Smelted together they made the alloy (metal mixture) bronze, used for tools, weapons, and utensils of the Bronze Age.

- **Iron ore was mined** and smelted about 3,000 years ago during the Iron Age. Much harder than bronze, iron was used to make tools, weapons, ornaments, and vessels.

- **About 95 percent** of all the metal in use in the world today is iron —some two tons for every person.

- **Around 98 percent** of all this iron is mixed with carbon, as charcoal or coke, during smelting to make the alloys called steels, which are stronger than pure iron.

- **Steels are used** in buildings, vehicles, railroads, household appliances, and all kinds of machines.

- **Brazil produces** the most iron, Australia is second.

- **The largest iron ore mine** in the world is the Kiruna Mine in Lapland, which is 3,428 ft deep.

- **About 43,000 years ago** early miners extracted the iron-containing pigment red ocher, which they used to decorate their bodies, from Lion Cave, Swaziland in Africa.

Gems and jewels

- **Gemstones or jewels** are either precious, such as diamond and ruby, or semi-precious, such as garnet and amethyst, depending on their rarity.

- **Most are mineral crystals** but some, like pearls, amber, and jet, come from plant or animal sources.

- **Gems are graded** for value by their color, clarity, and carats (weight). Color and clarity depend on the arrangement of the crystal's tiniest particles, atoms and molecules, and any traces or impurities.

- **Corundum contains** aluminum and oxygen. With traces of chromium it becomes red ruby, if iron and titanium are involved it becomes blue sapphire.

- **Most opaque gemstones** are polished into dome shapes called cabochons to show off their color and any internal patterning.

- **Transparent gems** like diamonds have flat faces or facets cut onto their surfaces, which allow the light to sparkle and "twinkle" as it passes through and reflects.

- **Most gemstones** formed millions of years ago deep in the Earth's crust where they slowly crystallized under great pressure and temperature.

◄ Clear sparkling diamond and black opaque coal are both forms of the same chemical element, carbon. In diamond the smallest particles, atoms, are arranged in boxlike shapes, in coal they are at random.

- **The crystals were brought** near to the surface by the slow churning of hot, melted magma, and they are mined from the rocks in which they formed.

- **The largest gem-quality diamond** ever found weighed 3,106.75 carats or 22 oz. It was found in 1905 in South Africa.

- **The second-biggest** man-made hole in the world is the Mir diamond mine in Russia, at 4,000 ft across and 1,700 ft deep.

- **The amount of diamonds** mined each year are worth an estimated $10.6 million. One-third are gem quality, the rest are industrial diamonds for drilling, grinding, and cutting.

▼ *Gems obtained from the now inactive Mir diamond mine, next to the city of Mirny, were contained in a rock called kimberlite. This was formed very deep under gigantic pressure, in a feature called a "diamond pipe."*

Fossil fuels

▼ Coal is usually taken by railroad to the nearest port and then loaded into bulk carrier cargo ships, as here at Nakhodka, southeast Russia.

● **Fossil fuels** were formed tens to hundreds of millions of years ago, when dead plants and animals were buried and subjected to very high pressures and temperatures.

DID YOU KNOW?

At the present rate of use, oil is estimated to run out by 2150, coal by 2250, and natural gas by 2300.

- **Natural gas** and petroleum (oil) were formed from planktonic sea organisms that sank to the seabed and rotted anaerobically, that is, without oxygen.

- **Coal and some natural gas** were formed from swampy forest plants that became buried, especially in the Carboniferous Period about 300 million years ago.

- **Fossil fuels** contain mostly carbon, which combines with oxygen when burned to form carbon dioxide (a greenhouse gas) and release heat energy.

- **Burning fossil fuels** is thought to be the main cause of global warming because it produces greenhouse gases.

- **Burning coal** also produces sulfuric, carbonic, and nitric acids that make acid rain, and a mixture of carbon particles and water vapor called smog.

- **Worldwide**, about 17 million tons of coal, 460 million cubic ft of oil and 3000 billion meters of natural gas are mined every day.

- **Coal is usually dug up** from holes in the ground, either from deep underground mines or from surface open cast mines.

- **Oil rigs drill down** to oil-bearing rocks and pump the oil up. A similar process is used to extract gas, often from the same rocks.

- **The deepest oil and gas wells** go down more than 39,000 ft below the surface.

Renewable energy

- **About 16 percent** of today's energy comes from renewable or sustainable sources, which will not run out in the foreseeable future. Also most do not produce greenhouse gases.

- **Most renewable energy sources**, such as wind, solar, hydroelectricity, and biofuels, harness the Sun's energy in some form. Geothermal energy comes from heat within the Earth.

- **Renewable energy sources** will not affect future generations. They also free nations from depending on other countries for energy, for example, buying coal and oil.

◀ Large wind turbines have a tower or pylon 230–330 ft tall and rotor blades 130–165 ft long. Large groups, called wind farms or arrays, have hundreds of turbines.

● **The use of wind turbines** is growing by 30 percent worldwide every year. A large turbine can generate enough electricity for 10,000 homes.

● **Brazil produces** the most biofuel in the world, ethanol, made from sugarcane. It powers 20 percent of the vehicles on its roads.

● **Before fossil fuels**, everyone used renewables. Firewood was probably the oldest, and wind was used by the Egyptians to power their boats 7,000 years ago.

● **Modern wind turbines** are very efficient. In theory they could provide as much as 40 times the total electricity used today.

● **The biggest hydroelectric power station** in the world is the Three Gorges Dam on the Yangtze River in China. It produces 20 times as much power as the Hoover Dam in the U.S.A.

● **About 450 nuclear power stations** produce some 13 percent of all the electricity in the world. Nuclear power is sustainable and does not emit greenhouse gases but it leaves radioactive waste that last for thousands of years.

● **Solar or photovoltaic cells** were invented in the 1880s. They turned just a few percent of the Sun's light energy into electrical energy.

● **Modern solar cells** convert 16–24 percent and the newest research versions reach 45 percent.

DID YOU KNOW?
To provide all the electricity used in the world, solar panels would have to cover an area one and a half times the size of the U.K.

Freshwater

● **The water cycle** involves the Sun's heat evaporating freshwater from the surface of both land and sea into water vapor, which cools and condenses into clouds and falls as rain or snow.

● **The rain or snow collects** in ponds, rivers, and underground aquifers. Much exists frozen as ice, as in Antarctica.

● **Less than 3 percent** of all the water on Earth is freshwater.

● **Only 0.1 percent** of this 3 percent (0.003 percent of the total) is clean, the rest is either salty or polluted by runoff of agricultural chemicals, factory waste, or sewage.

▼ *During dry spells, vast amounts of water are needed to irrigate crops such as corn. In some areas this use is not sustainable.*

- **Clean freshwater** is used for drinking and irrigation, washing, cooking, and sanitation.

- **Freshwater** is also used for factory processes, hydroelectricity, fishing, sailing, and watersports like waterskiing and whitewater rafting.

- **The thirstiest crops** are sugar cane and bananas, which need around 78 in of rain per year.

- **A dairy cow** needs 15–18 gallons of water daily.

- **As the human population increases**, water reserves shrink. About one-tenth of all people on Earth do not have free access to clean water.

- **The greatest reserves** of freshwater, such as the Great Lakes in North America and Africa, are usually far from the places which need the most water, for example India and China.

- **Water is becoming** an ever more valuable resource, and where rivers pass through several countries, such as the Nile in Africa, conflicts and even wars can arise if any nation interrupts the flow.

- **Aquifers are vast** but non-renewable sources of freshwater trapped in underground rocks, accessed by digging wells.

- **Huge untapped aquifers** containing 120,000 cu mi of water have been found under the continental shelves of Australia, China, North America, and South Africa.

Farming the land

- **Farming the land** to produce food, and fibers such as cotton and wool, began about 10,000 years ago. It was a major step leading to the great civilizations such as ancient Egypt and Mesopotamia.

- **Cereals or grains** (grasses) such as rice, corn, and wheat are the most important crops. More than 2.3 billion tons of cereals are grown every year, accounting for 60 percent of all the food humans eat.

- **Sugarcane is also a cereal**, grown throughout the tropics. It is the biggest crop in the world by weight, over 1.8 billion tons are grown per year.

- **Sugarcane and corn** are not just grown for food, they are also grown to produce biofuel. In 2013, almost 3 percent of the world's road transport was fueled by bioethanol or biodiesel, which give off less greenhouse gases than petrol.

- **Animals are farmed** for meat, milk, leather, and wool, and also used as working animals. Almost 20 percent of the Earth's land surface is used for animal production. Farm animals are also a huge source of greenhouse gases, as methane from their digestion.

- **Sheep cannot molt** their coats so they must be sheared or clipped every year. China has the largest "national flock" of 135 million sheep.

- **Poultry farms** produce turkeys and ducks but mostly chickens —more than 50 billion every year, as well as 1,200 billion eggs weighing 70 million tons.

- **Farm animals and plants** have always been improved by artificial selection for higher yields, disease resistance, or coping with harsh climates. Now scientists also use genetic engineering.

▲ Farming technology includes machines for plowing and harvesting, for milking, and also for chemical processes to produce fertilizer such as ammonium nitrate.

- **The "Green Revolution"** of the 1940s–60s with improved crop varieties, pesticides, fertilizers, machinery, and irrigation allowed one billion extra people to be fed.

- **Up to 70 percent** of all freshwater use in the world is for farming.

- **Irrigation systems** using pumps, pipes, canals, ditches, hoses, and sprayers bring water to crops when rain does not fall, even in the desert.

311

Exploiting the seas

- **Around 50 percent** of all the people in the world live within 50 mi of the sea. Many depend on it in some way, from fishing and boat-building to port workers and coastal tourism.

- **Overfishing has lead** to the collapse of some fisheries, and anchovy and sardine fisheries may never recover. Fish such as salmon are now farmed.

- **Most of the goods** traded around the world are moved on the seas by ships, especially container ships and oil tankers.

- **Ships and boats** are also important for travel and tourism. Cruising is a rapidly growing holiday occupation, while angling and other water-based sports have also become more popular.

- **Sand and gravel** for building are dredged from the shallow seabed. However the technology for mining minerals such as manganese, copper, nickel, iron, cobalt, and diamonds in deep water is usually still too expensive.

- **Drilling for oil** and natural gas beneath the seabed began in the 1890s. Oil and gas can now be extracted from deeper than 32,800 ft beneath the sea's surface.

- **Waves, tides, and variations** in salinity and temperature are all potential sources of renewable energy, but research into finding ways to harness them is slow, often due to the destructive power of storms.

- **No one knows** what new medicines may be found in the vast numbers of organisms in the oceans. An anti-HIV drug from a sponge and a pain-relief drug from coral are just two being tested by scientists.

- **The ocean is the main carbon sink** (storage reservoir), removing carbon dioxide from the atmosphere, and ocean plankton produces much of the oxygen we need to breathe.

- **The seas are being polluted** in many ways. The Great Pacific Garbage Patch is an area of floating debris and particles, many from plastics, that covers an area perhaps almost as big as Europe.

▼ The valuable metal manganese may be "harvested" as lumps called nodules on the deep seabed. These form as manganese minerals dissolved in sea water solidify onto an object such as a shell, growing just one millimeter every 100,000 years.

DID YOU KNOW?
Some 90 million tons of sea fish and shellfish are caught every year, which represents 15 percent of all the protein eaten in the world.

Living Earth

Ecosystems

- **An ecosystem** is a community of living things interacting with each other and their surroundings. Anything from a piece of rotting wood to a swamp can be an ecosystem.

- **When vegetation colonizes** an area, often the first plants to grow are small and simple, such as mosses and lichens. These plants stabilize the soil so that bigger, more complex plants can grow. This is called vegetation succession.

- **Rain forest ecosystems** cover only 8 percent of the world's land, yet they include 40 percent of all plant and animal species.

- **The food chain** is the route linking living things that feed upon each other. The eating interaction between them is rarely simple, so ecologists talk of food "webs."

- **Green plants** are autotrophs (producers)—they make their own food.

- **Animals are heterotrophs** (consumers)—they eat other living things.

- **Primary consumers** are herbivore animals (plant-eaters). Secondary consumers are either carnivores (meat-eaters), or omnivores (animals that eat plants and meat).

DID YOU KNOW?

Food chains on land usually have 3–4 links. Those in the ocean have as many as 8–10 links.

● **Decomposers or detritivores** include insects, millipedes, worms, bacteria, and fungi. They feed on the remains of dead plants and animals. In turn they prepare the soil for plants and help supply them with nutrients.

● **Every living thing** has its own unique role in the living community, called its niche. Because living species are so interdependent, a problem with one can affect the entire community.

▶ *All the living things in a natural ecosystem, such as a woodland, are dependent on each other.*

317

Biomes

- **A biome is a community** of plants and animals adapted to similar conditions in certain parts of the world. They are also known as "major life zones."

- **The vegetation or plant life** that grows in a region is mainly influenced by the soil and climate. In turn, plants greatly influence animal life.

- **Since vegetation is closely linked** to climate, biomes usually correspond to climate zones.

- **Major biome types** include tundra, boreal (cold) coniferous forests, temperate grasslands (prairies and steppes), savannas (tropical grasslands), tropical rain forests, deserts, wetlands, and mountains.

- **Most types of biome** occur across several continents.

- **Many plants and animals** have features that make them especially suited to a particular biome.

▶ Each different biome has its own range of creatures and plants.

Desert

- **Human activity** is putting many plant and animal species in danger. Forest habitats are lost as trees are cleared for farms and building. Habitats may also be poisoned by pesticides and other pollutants.

- **More than 40 percent** of all plant and animal species are at risk of extinction.

- **A recent survey** by an international team of scientists showed that less than 5 percent of the world's oceans are now entirely undamaged by human activity.

- **The fastest-growing biomes** are deserts due to desertification, and the urban "biome" of towns and cities, roads and parks, factories, and shopping malls.

- **By 2009** more people lived in urban environments than in rural or countryside ones.

Savanna

Rain forest

Biodiversity

- **Biodiversity means the variety**, or richness, of living things in a given environment or ecosystem. This is often measured as the numbers of different species, although other measures exist too.

- **The greatest biodiversity**, on land and in the oceans, occurs in warm climates, particularly between the Tropics near the Equator.

- **Tropical rain forests** (on land) and coral reefs (in the ocean) have the greatest numbers of species and so the greatest biodiversity.

- **Moving away from the Equator** toward the cooler temperate and polar regions, conditions for survival become more difficult and there is less biodiversity.

▼ *The Galapagos Islands, almost on the Equator in the Pacific Ocean, have great biodiversity including their unique giant tortoises.*

- **There are also biodiversity "hotspots"** around the world which are particularly rich in endemic plants and animals (those that occur nowhere else), such as Madagascar.

> **DID YOU KNOW?**
> A tropical rain forest may contain 1,000 different species of plants and animals in a tennis-court-sized area.

- **Colombia, in South America**, is a biodiversity "hot spot." It is home to 10 percent of all mammal species in the world, 14 percent of amphibians, and 18 percent of birds.

- **Biodiversity tends to increase** through time as living things evolve to exploit every tiny detail of their habitats.

- **The fossil record** in rocks shows increases and decreases in biodiversity over the Earth's history.

- **Past biodiversity** has been severely reduced by mass extinctions.

- **The worst mass extinction** occurred at the end of the Permian Period, 252 million years ago. Ninety percent of land species and 70 percent of marine species died out.

- **There are about 9 million species** alive today, but current rates of extinction are greater than ever, perhaps 100 times higher than even the Permian mass extinction.

Tropical forests

- **Tropical forests** grow in a band around the Earth between the Tropics of Cancer and Capricorn, either side of the Equator.

- **Average temperatures** of 68–86°F and as much as 33 ft of rainfall every year are perfect conditions for up to half of all the Earth's species of plants and animals.

- **The four main types** of tropical forest are: lowland equatorial evergreen rain forest, moist deciduous seasonal forest, montane or cloud rain forest, and flooded forest.

- **The forests grow in layers**: the forest floor, the dark understory, the rooflike canopy, and the emergent layer of huge trees that tower above everything else.

- **The forest floor** is in perpetual shade so few plants grow except around the edges and clearings. There is space for large animals such as tapirs, rhinoceros, and even elephants.

- **This is where nutrients** in fallen leaves or dead bodies are recycled by bacteria, fungi, and invertebrate decomposers such as millipedes and worms.

- **The understory** consists of tree trunks, saplings, and climbing vines growing up toward the light. Birds, snakes, frogs, and small mammals such as rats live here.

- **The canopy**, usually 98–130 ft above the ground, consists of a dense layer of branches, leaves, flowers and fruits, vines, and epiphytes (plants growing on other plants for support) such as bromeliads.

- **Millions of animals** live in the canopy, from insects and spiders, to birds such as parrots and toucans, and mammals such as monkeys, apes, and sloths.

- **Many emergent trees** reach heights of 180 ft, and some even grow to 260 ft. They are inhabited by large birds of prey, bats, and monkeys.

◄ *Red-bellied macaws are among the many colorful birds of the Amazon rainforest, feeding on year-round supplies of fruits and seeds.*

Temperate forests

- **The climate** to the north or south of the tropics consists of warm summers, cool winters, and lots of rainfall—temperate conditions.

- **In these cooler climates**, broadleaved trees lose their leaves at the beginning of winter—these are temperate deciduous forests.

- **Examples of trees** that grow in these forests are oak, maple, beech, chestnut, hickory, basswood, linden, walnut, sweet gum, and elm.

- **Forest life** largely comes to a halt in winter. But as soon as leaves appear in spring, they are eaten by hungry insects that come out of hiding or hatch from eggs. Birds then return to eat the insects.

- **The understory plants** in deciduous forests grow rapidly in spring. They flower and set seed, taking advantage of a brief period before the tree leaves cover them in shade.

- **The forest floor** is covered with fallen leaves and branches that support a community of detritivores, which eat dead and dying material, such as fungi, worms, and woodlice.

- **Temperate coniferous** or needleleaf forests grow mainly in the colder climates of the Northern Hemisphere. Pines, firs, and hemlocks are examples of these trees.

- **The highest biomass** (weight of biological material) is found in the temperate coniferous rain forests of North America, where the massive redwood trees grow.

- **Some of the animals** that live in temperate forests around the world are wolves, deer, bears, giant pandas, squirrels, and koalas.

- **As the Earth is divided** into forest zones, so are mountains, by height. Broadleaved trees tend to grow near the base and conifers higher up toward the tree line.

▼ *Fall, in Vermont, U.S.A. The leaves of trees turn to brilliant reds, pinks, oranges, and yellows. Trees withdraw nutrients from them into the trunks and branches, and prepare for the leaves to fall.*

325

Boreal forests

- **Boreal forest, or taiga**, consists of coniferous trees such as pines, firs, spruces, and larches, which grow in cold subarctic climates.

- **This type of habitat** is the largest land biome in the world. It stretches right across the northern parts of North America, northern Europe, and Asia.

▼ In the coldest part of winter, the American marten shelters in a burrow or cave under the snow for days at a time. Here it is much warmer than out in the open air.

DID YOU KNOW?

Boreal forests, including taiga, make up almost one-third of all of the Earth's forested areas.

- **There are long winters** when temperatures can be as low as −13°F, and a short summer growing season, which in some places is only two or three months long.

- **The more southerly taiga** consists of closed-canopy forest where the trees grow close together and only moss can survive on the dark forest floor.

- **Further north** there is lichen woodland. The cold often stunts the trees and they grow farther apart, with lichens in the light spaces between them.

- **Conifer trees** are adapted to cold conditions. Their narrow, waxy leaves prevent water loss and snow slides easily off their sloping branches.

- **Few larger animals** can thrive in the harsh conditions, but there are thousands of species of insect that overwinter as eggs or as larvae (in their early stages of development).

- **Animals in these forests** need ways of avoiding the long harsh winters. Bears, for example, stay in the forest and hibernate.

- **Larger animals** such as moose and elk also stay put. They grow thick winter coats and forage for food under snow during winter.

- **Most birds leave**, or migrate, in the fall, returning next spring to take advantage of the insects.

Polar and mountain life

- **The areas of the Earth** around the North and South Poles are covered with ice all year. The climate is very cold, windy, and dry.

- **The usually frozen ground** is called permafrost. If the surface melts in summer then algae, fungi, mosses, and some grasses and herbs can grow.

- **Some small invertebrates** such as spiders and mites manage to live in polar climates, and mosquitoes reach plague proportions where the permafrost thaws.

- **Larger animals** include Arctic foxes and hares, wolves, caribou, and musk oxen. The snowy owl, unusually for a bird, has a thick layer of fat to keep it warm all year.

- **The largest land-based predator** in the Arctic is the polar bear, which hunts seals from ice floes.

- **Conditions in Antarctica** are more severe than in the Arctic. Emperor penguins are the only larger animals that live there on land, and even raise their chicks on the ice.

- **The polar seas** are rich in nutrients and squid, fish such as cod and herring, and krill thrive. Larger animals such sperm whales and other great whales migrate to these waters to feed.

- **Arctic birds** include kittiwakes, fulmars, little auks, and murres. They breed here in the spring when the ice starts to break up.

- **The region between the Arctic** and the boreal forest is called tundra. It has few or no trees, mainly algae, lichens, mosses, grasses, and a few stunted willows.

- **Mountains are cold**, windy places where only certain animals can survive. These include agile hunters such as pumas and snow leopards, and nimble grazers such as mountain goats, yaks, ibex, and chamois.

▼ Bighorn sheep leap among crags of the Rocky Mountains with ease, their non-slip hooves gripping even icy rocks.

Grasslands

- **About one-fifth** of the land on Earth is covered by grasslands, found on every continent except Antarctica.

- **Grasslands are called prairie** in North America, savanna in Africa and Australia, pampas in South America, and steppe in Eurasia.

- **Grass plants** have adapted to being eaten by animals, and their delicate growing point is hidden and protected deep down among the leaves or blades.

- **Trees do not grow** because saplings are eaten by grazing animals.

- **Periods of drought**, and wildfires caused by lightning strikes in late summer when the grass is dry, inhibit many other plants.

- **Grazing animals** include wildebeest, bison, gazelles, and prairie dogs, which provide food for large predators such as lions, cheetahs, coyotes, and wolves.

▼ *Wildebeest and zebra graze the African savanna. Their sharp eyesight, keen sense of smell, and acute hearing pick up on predators such as lions and hyenas. With no hiding place, escape on long legs is the main strategy.*

- **Prairie and steppe** are temperate grasslands, growing just outside the tropics. They have hot summers and cold winters, rain in spring, and drought in fall and winter.

- **The deep, rich soil** of the prairie and steppe is full of invertebrates such as nematodes and earthworms.

- **Savanna grassland**, also called tropical grassland, is found in hotter climates, especially Africa. It has a few small trees and shrubs and thin, free-draining soil.

- **Grasslands have low biodiversity** but high biomass, that is, there are relatively a few species but very high numbers of each.

- **There are hundreds of species of grasses**, and wildflowers that grow among them include asters, sunflowers, clovers, wild indigos, goldenrod, and blazing stars.

Life in dry lands

- **Arid, semi-arid, or desert regions** are those where the amount of rainfall is less than the amount of evaporation. They are usually very hot in the day and cold at night.

- **This kind of biome** covers 20 percent of the Earth's land surface. The largest is the Sahara in Africa, which covers more than 3 million sq mi.

- **Plants and animals** that live here must be highly specialized to find enough water and food, and prevent overheating.

- **Desert plants** have various ways of coping with little water. Some have very deep roots, and some have roots that cover a wide area just under the ground surface to soak up any rain that does fall.

- **Many plants** have small, waxy leaves, or none at all, and contain toxic chemicals or grow thorns to stop animals eating them.

- **Cacti are typical desert plants**. They have leaves modified as spines, a thick waxy coating that prevents water loss, and green stems that can expand when water is available.

- **Many wildflowers** that live in dry regions go through their life cycle very quickly. They germinate, grow, flower, and make seeds within a few days after rain.

- **Small rodents** such as mice, and reptiles such as lizards and snakes, live in deserts, along with hardy invertebrates such as beetles and scorpions. Most live in cool burrows by day and get enough water from their food.

- **Camels have many adaptations** for desert life. They can drink huge amounts of water after going without it for days, their wide feet do not sink in the sand, and their long eyelashes keep sand out of their eyes.

- **The fennec fox** has huge ears so it can hear the tiniest rustle of a small creature such as a beetle when hunting at night. Its ears also make good radiators to help get rid of excess heat.

▼ In the desert and scrub of southwest North America, saguaro cacti up to 80 ft tall provide resting places for birds and bats.

Wetland and freshwater

- **Wetlands are areas** that are permanently or regularly under fresh or salty water, such as ponds, rivers, bogs, marshes, estuaries, and mangrove swamps.

- **They are found** on all continents except Antarctica. The Amazon river basin is the largest wetland in the world.

- **There are four types** of wetland plants: submerged plants such as eelgrass, floating plants such as duckweed, emergent plants such as water lily, and those that grow at the water's edges or margins such as mangrove trees.

- **Wetlands have very high biodiversity**, or numbers of different species. Animals include invertebrates, fish, amphibians, reptiles, birds, and mammals.

- **Invertebrates include** crabs, crayfish, and other crustaceans, pond skaters, diving beetles, and other insects, worms and leeches, and water snails.

- **Many insects**, such as dragonflies, and all amphibians, such as toads and newts, live in water for part of their life cycle, as nymphs or tadpoles.

- **Freshwater fish**, from minnows to huge carp, perch and pike, make up 30 percent of all known fish species.

- **Frogs are used as indicator species** to gauge the health of the wetland since they are very sensitive to any toxic chemicals that may be polluting the water.

- **Reptiles that live** in and around water include turtles, alligators, crocodiles, and snakes such as the giant anaconda.

- **Huge numbers of birds** depend on wetlands. There are water birds such as grebes and ducks, wading birds like avocets and curlews, predatory birds such as ospreys and bald eagles, and migratory birds such as swans and geese.

- **Millions of migrating birds** stop at wetlands on their long journeys to rest, refuel, and socialize.

- **Large wetland mammals** include capybara, hippopotamus, and lechwe antelopes.

▼ *American alligators inhabit the swamps of the southeast U.S.A. Growing to 15 ft in length and 880 lb, they are—like crocodilians elsewhere—the largest predators of wetland habitats.*

Freshwater life

- **The African lungfish** can survive if its pond dries up. It burrows into the mud at the bottom, secretes a protective mucus cocoon, and goes to sleep until the water returns.

- **Fish that live in lakes** and rivers in caves, such as the Mexican cave characin, lose their sight because they live in total darkness.

- **The duckbilled platypus** is a strange egg-laying mammal of muddy Australian waterways. It finds its prey by sensing electrical fields with its leathery "beak."

- **Waterbuck**, sitatunga, and the water chevrotain are African hoofed animals (ungulates) that live only near water.

▼ *The northern pike, up to 5 ft long, is the top predator of many cooler rivers and lakes. It can grab sizeable fish whole in its huge mouth.*

- **Otters and mink** are predators of waterways. Superb swimmers, they are highly adapted for catching and feeding on fish.

- **Beavers create** their own freshwater habitats by felling trees to dam rivers and create ponds, so they can build homes called lodges to protect their young.

- **The larva of the drone fly** is called a rat-tailed maggot. It lives in stagnant fresh water and has a long snorkel-like tail to breath air from the surface.

- **There are freshwater species** of many sea shellfish such as mussels and clams. Freshwater mussels even produce pearls, like those of marine oysters.

- **Pond-edge plants** include reeds, sedges, and rushes. Waterlilies grow in deeper water and water hyacinth floats on the surface.

- **The water hyacinth**, originally from South America, grows faster than almost any other plant. It can cover the water's surface to a depth of 3.3 ft, smothering all other life. It has become a major problem around the world.

337

Coastal habitats

- **Coastal habitats** start at the high tide mark of sandy and rocky shores, and go to the edge of the shallow zone called the continental shelf. They include estuaries, fjords, and reefs.

- **These habitats form** about 7 percent of the ocean surface, but most marine animals live here and they represent 30 percent of ocean productivity (growth of living matter).

- **This productivity** is based on primary producers. These are algae (seaweeds) and tiny floating phytoplankton, which use sunlight to grow and become food for all coastal creatures.

- **Estuaries**, where rivers meet the sea, are challenging but highly productive habitats. The water is brackish—a mixture of salt and fresh that changes twice a day as tides come and go.

▼ In this estuary in Taiwan, East Asia, mudskipper fish and fiddler crabs cope with ever-changing conditions of tides, fresh river water, the salty sea, waves, drying winds, tropical heat, and intense downpours.

- **Estuaries provide a nursery** for many baby fish, and the thick silty mud provides a nutritious home for burrowing shellfish.

- **Coral reefs form** in shallow water where tiny algae that live within the coral polyp's body trap energy in sunlight by the process of photosynthesis.

- **Sandy shores**, or beaches, are continuously shifted by the tides and waves. Sea lions and seals, gulls and terns, shellfish such as clams and periwinkles, shrimp, starfish, sea urchins, and many worms live here. Turtles lay their eggs in the sand every year.

- **Wave-washed shingle** or pebble beaches are harsh and unstable habitats. Seaweeds cannot grow on them. But there are invertebrates such as isopods that live on bits and pieces cast up by the waves. Birds such as the plover nest here.

- **Rocky shores** are more stable, although the waves erode the rocks eventually. Barnacles and mussels stick to the rock surface between high and low tide lines. Sea anemones, limpets, crabs, shrimp, and small fish live in the rock pools.

- **Kelp is long, strap-shaped seaweed** that forms underwater forests below the low tide mark in cooler seas. They can grow 12–24 in in one day.

Life in the open ocean

- **The open ocean** starts at the edge of the continental shelf where the seabed plunges suddenly down the continental slope to greater depths.

- **The open ocean** is divided into three zones: the benthic zone is at the bottom, above that is the demersal zone, and the pelagic zone is uppermost.

- **Sunlight only passes** about 650–1,600 ft down through seawater, so plant photosynthesis can only occur near the surface.

- **Tiny floating phytoplankton** are the main plant life of the open ocean. They drift with the currents and form the basis of all open ocean animal life.

- **Zooplankton feed** on the phytoplankton. Zooplankton are made up of the larvae (young forms) of fish, starfish, crabs, and many other animals.

▼ dolphins are fast, agile predators of the open ocean, speeding along at 37 mph to catch fish and squid.

- **Shrimplike animals** called krill feed on all forms of plankton and are among the most numerous animals on Earth.

- **A "rain" of dead animals** and excrement, called marine snow, sinks down from the upper layers and feeds the creatures of the dark layers below.

- **Warm currents** and differences in salinity cause upwellings of deep water that bring nutrients to the upper layers, especially near the edge of the continental shelf.

- **Many creatures** feed on these circulating banquets by straining out small edible items—filter-feeders such as whale sharks, corals, bristleworms, and sponges.

- **Filter-feeding fish** such as herring attract large predatory fish such as tuna and sharks, and also seabirds such as terns and shearwaters above the water.

- **The largest filter-feeder** in the open ocean is the largest creature in the world. The blue whale, which grows up to 98 ft long and weighs up to 150 tons, strains about 40 million krill from the water every day in summer.

Deep sea life

Deep sea begins where light fades and plants can no longer carry out photosynthesis. Depending on water conditions, below 1,600–3,300 ft is permanent blackness.

The pressure at the bottom of the deep sea may be 1,000 times that at the surface, and temperatures are just 35–40°F— except at hydrothermal vents where the water may reach up to 1,200°F.

There are only three sources of food in the depths: "marine snow" of dead animals and excrement falling from the surface waters, hydrothermal vents, and whale carcasses called "whale falls."

Marine snow takes several weeks to fall to the deep seabed, and much is consumed by filter-feeders on the way down.

What is left of marine snow forms a nutritious ooze on the deep ocean floor. Worms, sea snails, shrimp, giant isopods, starfish, sea cucumbers, and sea urchins burrow through this ooze, feeding as they go.

When a dead whale sinks to the bottom, small creatures arrive within minutes, followed by hagfish, sleeper sharks and shrimps. It may take 50–100 years for the carcass to disappear

- **Creatures called bone-eating snotworms** or zombie worms finish off a whale fall by consuming even the bones.

- **The deepest sea life** is in deep-sea trenches where, despite the extreme crushing pressure, flatfish, shrimp, and jellyfish have been seen by research submersible craft.

- **Deep-sea creatures** that live in midwater never meet a hard surface and so their bodies can be any shape, and they don't need a skeleton as the water supports them.

- **Many deep-sea creatures** use bioluminescence, which allows them to make light. They then use it to find prey or a mate, or to avoid predators.

- **Bioluminescent animals** are hugely numerous and varied and include combjellies, sea pens, jellyfish, shrimp, and krill, squid, and many fish such as lanternfish, flashlight fish, hatchetfish and viperfish.

- **The largest invertebrates** in the world are the giant and the colossal squid. They live in the deep sea and grow to 46 ft long and weigh more than 1,100 lb.

◄ *Fierce-looking fangtooth fish can swim to depths of around 16,400 ft, into the Abyssal Zone, when they follow their prey.*

Earth in danger

Global warming

- **The general increase** in average temperatures around the world is called global warming. This increase has been almost 1.5°F over the last 100 years.

- **Most scientists now think** that global warming is caused mainly by human activities, which have resulted in an increase over and above the Earth's natural greenhouse effect.

- **The greenhouse effect** is how certain gases in the air—notably carbon dioxide, ozone, and methane—trap some of the Sun's warmth, like the glass in a greenhouse.

- **The Earth is kept warm** by the natural greenhouse effect, at about 59°F on average, rather 5°F without a greenhouse effect.

- **If too much heat** is trapped, the Earth may become hotter.

- **Many experts expect** a 3.5°F rise in average temperatures over the next 100 years.

▶ If the world continues to warm, in winter, the Arctic ice cap may shrink to half the size it is today.

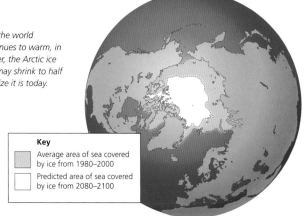

Key
- Average area of sea covered by ice from 1980–2000
- Predicted area of sea covered by ice from 2080–2100

● **Humans boost** the greenhouse effect by burning fossil fuels, such as coal and oil, which produce carbon dioxide.

● **Emissions** of the greenhouse gas methane from the world's cattle and other farm livestock add to global warming.

● **Global warming** makes the oceans increase in volume by expansion. It may also melt much of the polar ice caps, flooding large areas of low-lying land with major cities around the world.

● **By trapping more energy** inside the atmosphere, global warming is also predicted to bring climate change and stormier, more extreme weather.

▶ In Greenland, scientists drill deep into the ice and the ground below it to extract core samples, which hold clues about past changes in weather and climate.

Climate change

- **The world's climate** has changed through time, getting warmer, colder, wetter, or drier. There are many theories why this happens.

- **One way to see** how climate changed before weather records were kept is by growth rings in old trees. Wide rings show the good growth of a warm summer.

▼ Some changes in weather are linked to fluctuations in sunspots, dark spots on the Sun. They can cause stormy weather on Earth.

- **Another way of working out** past climate is to examine ancient sediments in lakes and seas for the remains of plants and animals that only thrive in certain conditions.

- **Another method** is to drill ice cores from glaciers and ice caps, where atmospheric gases, pollen, and dust from past ages are trapped.

- **One cause** of climate change may be shifts in the Earth's orientation to the Sun. These shifts are called Milankovitch cycles.

- **Among the Milankovitch cycles are:** the change in shape of the Earth's whole orbit around the Sun, in a 100,000-year cycle; the slight increase or decrease in the tilt of the Earth's axis as the Earth orbits the Sun, in a 40,000-year cycle; and the slight wobbling, or precession, of this axis as the planet spins, in a 21,000-year cycle.

- **Climate may be affected** by patches on the Sun called sunspots. These flare up and down every 11 years.

- **Climates may cool** when the air is filled with dust from volcanic eruptions.

- **Local climates may change** as continents drift. Antarctica was once in the tropics, while New York, U.S.A., once had a tropical desert climate.

- **Climates may get warmer** when levels of certain gases in the air increase, for example, with mass volcanic eruptions, and as is happening today with greenhouse gases.

Pollution

● **Air pollution comes mainly** from vehicle exhausts, waste burners, factories, power stations, and the burning of oil, coal, and gas in homes.

● **Pesticide crop sprays**, farm animals, mining, and heavy industries also contribute to air pollution.

● **Some pollutants**, such as soot and ash, are solid as tiny particles or particulates, but many more pollutants are gases.

● **Air pollution can spread** huge distances. For example, pesticide traces have been discovered in Antarctica where they have never been used.

● **Most fuels are made up** of chemicals called hydrocarbons (hydrogen and carbon). Any hydrocarbons left unburned in vehicle exhausts can react in sunlight to form toxic ozone at ground level.

- **When exhaust gases react** in sunlight to form ozone, they may create a harmful photochemical smog.

- **Air pollution** is probably a major cause of global warming and has harmed the ozone layer of the Earth's atmosphere, which protects the surface from the Sun's harmful ultraviolet rays.

- **Breathing polluted air** in big cities is thought to be as harmful as smoking 20 cigarettes a day.

▶ Most gas emissions from power station stacks or chimneys (upper right) are now filtered or "scrubbed" to reduce pollution.

Acid and ozone

▲ Many trees are killed by acid rain—rain that has become acidic due to air pollution, caused by burning fossil fuels.

- **All rain is slightly acidic**, but air pollution can turn rain into harmful, corrosive acid rain.

- **Acid rain forms** when sunlight makes sulfur dioxide and nitrogen oxide combine with oxygen and moisture in the air.

- **Sulfur dioxide** and nitrogen oxide come mainly from burning fossil fuels—coal, oil, and natural gas.

- **In some developed regions** such as Europe and North America, acid rain has been cut dramatically by government regulations to reduce sulfur dioxide emissions from power stations.

- **Acid rain damages** plants by taking nutrients from leaves and blocking the plants' uptake of nitrogen.

- **Twenty percent** of trees in Europe, and up to 50 percent of trees in some places, were damaged by acid rain.

- **Life on Earth** depends on the layer of ozone gas in the air, which shields the Earth from the Sun's ultraviolet (UV) rays.

- **In the early 1980s**, scientists noticed a loss of ozone over Antarctica, which peaked each spring.

- **The loss, depletion, or thinning** of ozone was caused mainly by manufactured gases, notably chlorofluorocarbons (CFCs).

- **Most CFCs were banned** in the 1990s, and the ozone layer is now mending naturally very slowly.

- **It may be** another 100 years before the ozone layer returns almost to its natural state.

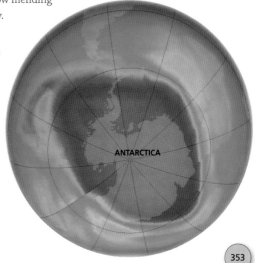

▶ The ozone thinned area (blue) is loss of ozone density over the South Pole. In recent years it has stabilized and may be fading.

ANTARCTICA

Drought and desertification

- **A drought is a long period** when there is too little rain.

- **During a drought** the soil dries out, streams stop flowing, groundwater levels of the water table sink, and plants die.

- **Deserts suffer** from permanent drought. Many tropical places have seasonal droughts, with long dry seasons.

- **Droughts are often accompanied** by high temperatures, which increase water loss through evaporation.

- **In North America**, the Great Drought of 1276–1299 destroyed the cities of the native civilizations of the southwest, called the ancient Pueblo culture, and cities were abandoned.

- **In the 1870s** severe drought in China killed crops and livestock, and an estimated nine million people died.

▼ *In severe droughts, even waterholes dry out, leaving nothing but cracked mud.*

 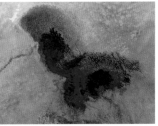

▲ *In 1973 Lake Chad (dark blue, above left) covered some 9,650 sq mi in dry North Africa. By 2007 repeated droughts, desertification, and use of water for irrigation had ravaged its extent (above right).*

- **Between 1931 and 1938**, drought reduced the Great Plains of the U.S.A. to a dustbowl, as the soil dried out and became windblown dust. Drought occurred there again from 1950 to 1954.

- **Desertification** is the spread of desert conditions into surrounding grassland. It is caused by climate changes and related human activities such as too many livestock and trying to grow too many crops.

- **Combined with increased numbers** of livestock and people, drought has put pressure on the Sahel region, south of the Sahara, in Africa, causing widespread desertification.

- **Drought has brought** repeated famine to the Sahel, especially Sudan and Ethiopia.

- **Drought in the Sahel** may be partly triggered by El Niño—a reversal of the ocean currents in the Pacific Ocean, off Peru, which happens every two to seven years.

DID YOU KNOW?

By the late 1990s Lake Chad in North Africa shrank to just 5 percent of its size in the 1970s, but it has since enlarged.

Overfarming

- **Overfarming, or overharvesting**, is when land is exploited to such an extent that it becomes unsuitable for further farming.

- **Intensive farming** is a way of producing as much food as possible in the minimum area possible, despite greater costs. It depends on increased use of machines and chemicals such as fertilizers and pesticides.

- **Factory farming** is the rearing of lots of animals in small enclosures, usually indoors. Factory farms produce 70 percent of the world's eggs, 75 percent of poultry, 40 percent of beef, and 50 percent of pork.

- **The amount of food** these animals eat is kept to a minimum optimum level, and their health is controled with medicines. However their welfare may be considered less important.

- **Land degradation** results from the removal of the plants that cover, protect, enrich, and bind the soil.

- **Overgrazing is allowing** too many animals (such as goats) to continuously eat pasture without allowing time for land to recover. This uses up the nutrients and exposes the soil to erosion.

- **Animals that graze** poor pasture lose condition, producing less milk or meat, and are less able to produce healthy offspring.

- **Land degradation** results eventually in desertification, where fertile land is turned to desert. It probably affects more than 3.8 million sq mi.

- **Poverty drives** land degradation because too many people have no choice but to try and produce more food from soil that has become exhausted.

- **Deforestation**, when trees are cut down to for timber, and to create fields for crops or livestock grazing, also degrades land.

- **Some surveys estimate** that two-thirds of all useable agricultural land has now been degraded.

▼ *Intensive farming involves plants (here in glasshouses) or animals raised in large numbers, packed closely together, in highly controled conditions of heat, light, air, water, and food—all using vast amounts of money, materials, and resources.*

357

Vanishing resources

- **Naturally occurring** resources are substances people use for energy, food, or manufacturing. Some are renewable, some are not.

- **Resources such as solar** and wind power and geothermal energy are not vanishing, they are naturally renewed.

- **Energy resources** such as coal and oil may last for only another one or two centuries at current rates of use.

▼ In Mato Grosso state, Brazil, workers install a tank to store clean water pumped up from a new well. Local surface water can be very polluted from mining and farming.

- **Poor farming practices** may mean food production cannot be maintained because the land, another natural resource, has lost its fertility.

- **Another food resource** at risk is fish and shellfish. Overfishing results in fish being unable to reproduce fast enough to recover numbers. The Atlantic cod fisheries collapsed by nine-tenths by the 1990s and show little signs of recovery.

- **Clean freshwater**, used for drinking, washing, watering animals, and irrigation, may eventually become very scarce. Up to one-quarter of the world's population may not have access to clean drinking water by 2050.

- **Clean air is a resource** that will not run out in the long term. But air can become so polluted, as in China, that it causes illnesses and prevents the growth of food plants.

- **Metals come from rocks** called ores that are found in the ground and which are not renewable. Iron, for example, may run out in 100–150 years at the rate it is currently used.

- **Phosphate is an essential element** that is used to make fertilizers for growing crops. There may only be enough for another 100–200 years.

- **Certain chemical elements** called rare earths, essential for today's electrical devices from smartphones to wind turbines, also have a limited supply, estimated at 250 years or less.

Threatened species

- **The International Union for Conservation of Nature**, IUCN, is the main organization that decides which species of animals and plants are in danger of dying out or becoming extinct.

- **More than 3,000 animals** and 2,600 plants—over 40 percent of all studied species—are on the Red Lists of Threatened Species.

▼ *Among the most endangered large animals are rhinos. There are fewer than 300 Sumatran rhinos left in the world. Listed by the IUCN as CR, critically endangered, they are probably—tragically—doomed to extinction.*

- **The IUCN classifies** species as extinct EX, extinct in the wild EW, critically endangered CR, endangered EN, vulnerable VU, near threatened NT, or of least concern LC.

- **Other categories include** data deficient DD, with not enough information to make a judgement, and NE, not evaluated as yet.

- **Wild animals have been hunted** for thousands of years, primarily for food, but also for fashionable furs, feathers, skins, scales, oils, traditional medicines, or aphrodisiac powders.

- **Some of these substances** and materials are worth large amounts of money in certain countries and therefore illegal poaching and trade are hard to prevent.

- **Whales were once hunted** for their oily blubber and meat. Now most countries abide by the global whaling ban and numbers are slowly increasing.

- **Climate change** and habitat loss are causing many animals to become endangered, reducing suitable areas where they can live.

- **The present loss of animals** and plants is called the Holocene Mass Extinction. Fossil records suggest it is the fastest extinction event ever.

- **It is estimated** that up to 40 percent of all the species on the planet could disappear in the next 100 years.

- **The extinctions now being caused** by human activity are called the biotic crisis. If past mass extinctions are a guide, Earth may take 5–10 million years to recover.

Conservation

- **Conservation aims to protect** natural and wild places and their animals and plants.

- **Captive breeding** is one way of keeping a species alive by providing it with a safe environment to reproduce until it can be released back into its habitat, if and when that habitat is safe.

- **There are many problems** with captive breeding such as possible inbreeding or animals not learning survival skills.

- **If there is no suitable habitat** left to reintroduce the animals produced, captive breeding cannot be successful.

- **Private farming** is a way of encouraging local people to protect species so they can profit from them by tourism or by supplying the approved pet trade.

- **Zoos and other breeding centers** swap the most distantly related animals of a species so they can breed together to ensure the greatest genetic variation.

- **Nature reserves**, where wild animals are protected along with their habitat, have been established all over the world on land and in the sea.

- **Some conservationists** are concerned that a lot of money is being spent on a few iconic animals. They suggest that sacrificing some of these might give lesser thought-about animals, plants, and fungi a chance to survive.

- **Some experts say** that it is more important to protect whole ecosystems—animals, plants, and the environment where they live—rather than individual species.

▲ *Part of successful conservation is knowledge of an animal's habits, movements, and foods. This hawksbill turtle is fitted with a transmitter that tracks its travels through the ocean while feeding, mating, and laying eggs.*

● **Others say that campaigns** to conserve a particular and famous animal in the wild, like the panda, gorilla, rhinoceros, or tiger, are more successful with the public than general campaigns and have the same end result.

● **The first medium-large mammal** announced as extinct in the Third Millennium was the baiji or Yangtze River dolphin.

Future Earth

- **There may be 9–10 billion people** by 2050. These will be mostly young people in developing countries and it will be difficult to produce enough food for all.

- **Humans may go extinct** due to global famine, new diseases, a nuclear accident, warfare, runaway greenhouse effect, essential resources running out, some unforeseen catastrophe, or a combination of these.

▼ *Many fanciful designs exist for future "vertical cities" with plants growing in "farmscrapers." However the resources needed and costs would be gigantic. The real future is likely to be much more "low-tech."*

- **The Earth may be affected** by extraterrestrial events such as a massive asteroid coming too close or even hitting it, or a nearby supernova (exploding star) engulfing it.

- **Looking farther ahead,** and depending on the current global warming crisis, Earth should slowly enter the next ice age in 50,000–100,000 years.

- **In 500 million years,** carbon dioxide levels may fall as the Sun becomes hotter and brighter, thereby stopping plant photosynthesis.

- **The Earth's tilt may alter** by as much as 90 degrees in the next billion years, drastically altering the seasons as oceans evaporate and tectonic plates grind to a halt.

- **By 2.3 billion years** the Earth will lose its magnetic "shield" that protects against solar radiation. Remaining life on Earth will end.

- **In 5–7 billion years** the Sun will expand as a red giant star and absorb the inner planets of Mercury, Venus—and Earth.

- **Such cosmic events** in the far-off future are likely impossible for humans to influence. But we can do huge amounts today to secure better centuries to come.

- **Renewable energies,** decreased waste, increased recycling, more efficient food production with healthier plant-based diets, conserving and sharing resources, habitats and wildlife—and, some argue, slower population growth—can all assist our planet Earth to support us in forthcoming generations.

Index

Index

Page numbers in **bold** refer to main subject entries; page numbers in *italics* refer to illustrations.

A

"aa" lava 78
Aboriginal people 207
absolute humidity 261
abyssal plains 178, *179*
acid rain 100, 104, 305, **352–353**
Aconcagua, Lake 149, 199
advection fog 265
Africa 54, 132, **136–137**, 208, 209
African tectonic plate 159, 204, 212
Afro-Eurasia 132
air moisture **260–261**
air pressure 67, **272–273**, 276
albedo 259
Aleutians *59*, 195
alligators *335*
alloys 301
Alpine Belt 195
Alps *68–69*, 134, **204–205**
altocumulus clouds 262
altostratus clouds 262
aluminum 33, 35, 297, 302
Amasia 55
Amazon rainforest 148, 230, 281, 323
Amazon River *106*, 119, 222, **224–225**, 334
amber 29, 35, 302
America 132, 150, 208, 209
American Cordillera 195, 201
amethysts 205, 302
Andes 67, 68, 69, 148, 149, 195, **198–199**
Angel Falls 115
animals,
    coastal **338–339**
    conservation **362–363**
animals (*cont.*)
    deep-sea **342–343**
    deserts 332, 333
    evolution 23
    farm 309, 310, 347, 356
    forests 322, 323, 324, 325, 327
    freshwater habitats 245, 334, **336–337**
    grasslands 330, 331
    indicator species 334
    marine 174, 175, 181, **340–343**
    mountains 67, *198*
    polar regions *145*, **328–329**
    threatened species **360–361**
    wetlands **334–335**
Annapurna, mountain 218
Antarctic Circle 149
Antarctic Circumpolar Current 163
Antarctic tectonic plate 159, 199
Antarctica 122, 132, 140, **144–145**, 150, 162, 177, 210–211, 250, 308, 328, 349, 353
anticlines 65
anticyclones 273
aphanitic rocks 38
Apostles *173*
Appalachian Mountains 146
aquifers 308, 309
Arabian Sea 217
Aral Sea 120, *121*
arches 172, *172*
Arctic 176, 328
Arctic Circle 146, 248
Arctic Cordillera 195
Arctic Ocean 146, 152, *153*, **156–157**
arêtes 127
argon 254
arroyos 123
Ascension Island 155, 212
Asia 55, 132, **138–139**
astatine 33
asteroids 13

Aswan Dam 227
Atacama Desert 183
Atacama Trench 187
athenosphere 52
Atlantic Ocean 58, 60, 152, *153*,
    154–155, 164, 184, 186, 191, 212
Atlas Mountains 137, 154, **208–209**
atmosphere 13, 32, **254–255**, 292
Aurora Australis 257
Aurora Borealis 256, 257
auroras *255*, **256–257**
Australia 25, 132, **140–141**, 142, 150, 158
Australian plate 206
autotrophs 316
avalanches 98
Azores 155, *155*, 212

B

Baffin Island 151
Baikal, Lake 120, 238, **242–243**
Baja California Desert 183
barchans 122, *122*
barometers 273
bars 173
basal slip 124
basal weathering front 100
basalt 37, 38, 51
Bay of Bengal 173, 234
Bay of Fundy 167
bays 171, *172*, 173
beach cusps 171
beaches **170–171**, 339
Beardmore Glacier 211
Benioff–Wadati zones 59
benthic zone 153, 340
bergschrund 124
berms 171
Bicaz Canyon 219
bighorn sheep *329*
bights 173
biodiversity **320–321**, 331, 334

biofuels 307, 310
bioluminescence 343
biomass 325, 331
biomes **318–319**
biotic crisis 361
black ice 270
black smokers **180–181**
blizzards **282–283**
Blue Mountains 207
Blue Nile 226, 227
blue whales 341
bogs 119, 334
Bølling-Allerød interstadial 27
Bolshoye Shchuchye, Lake 203
boreal forests 318, **326–327**
Borneo 151
Bouvet Island 151
brackish water 338
Brahmaputra River 197, 234
"breadcrust" bombs 79
breakers 168
breezes 276
The Burren *102*
buttes 123

C

cacti 332, *333*
calcium carbonate 104, 174
calderas 73, *75*, *238*, 239
Caledonian mountains 134
calving 176
Canadian Shield 147
Cancun *170–171*
canopy (forest) 323
canyons and gorges *99*, 103, 111, 147,
    214–219
captive breeding 362
carbon 33, 301, 305
carbon dioxide 27, 72, 100, 103, 254,
    305, 313, 346, 365
carbon monoxide 254

carbon sinks 313
carbonation 103
carbonic acid 103
Carboniferous Period 23, 305
Caribbean Sea 183, 190, 191
Caribbean tectonic plate 191
Carlsbad Caverns *105*
carnivores 316
Caspian Sea 120, 139, 203
cataracts *231*
caverns 103, 104, *105*
caves **104–105**, 172, 336
Cenozoic Era 22, 209
cereal crops 310
Chad, Lake 355
Challenger Deep 188, *189*
Chambeshi River 230
chlorofluorocarbons (CFCs) 353
chrons 22
cinder cones 74
cinders 79
Circum-Pacific Mountain System 195
cirques 125, *125*, 127
cirrocumulus clouds 262
cirrostratus clouds 262
cirrus clouds 262, 274, *275*
cliffs 172, *172*, *173*
climate 290
climate change 347, **348–349**, 361
climate optima 26
climate zones **290–291**, 318
clints 103
clouds 260, **262–263**, 266, 274, 275,
    278, 284
coal 23, 35, 36, 42, 201, 202, 203, 209,
    211, 296, 304, 305, 347, 350, 358
coasts,
    plants and animals **338–339**
    rocky **172–173**, 339
    sandy **170–171**, 339
cobalt 312
cold **128–129**, 270

cold fronts 274, 275, *275*, 284
Colombia 321
Colorado Plateau *98–99*, 215
Colorado River *99*, 111, 201, 214
Columbia River 201
composite volcanoes 72, 74, *75*
compression faults 62
condensation 260, 262, 264, 269, 308
condors *198*
Congo River 119, **230–231**
coniferous forests 138, 318, 325, 327
conservation **362–363**
continental climate 290
continental crust 51
Continental Divide 201
continental drift **54–55**, 349
continental rise 178
continental shelf 178, *179*, 338, 340
continental slope 178, *179*
continents 56
    definitions of **132–133**
    subcontinents 132
    supercontinents 54, 55, 132, 140
    see also individual index entries
contrails 262
convection currents 48, 56
convexo-concave slopes 109
conyphytons 28
copper 21, 34, 35, 181, 199, 201, 203,
    205, 209, 296, 297, 301, 312
Copper Canyon 218
coral islands 159, 161
coral polyps 174, *175*
coral reefs 27, 143, 161, **174–175**,
    320, 339
cordilleras 195
corries 127
corundum 302
coves 173
Cox's Bazaar 171
crater lakes 73
Cretaceous Period 23

Cretaceous Thermal Optimum 27
crevasses 124, *125*
crystals 34, 38, 46, 205, 302–303
  recrystallization 41
cumulonimbus clouds 262, 266, *275*, 278
cumulus clouds 262, 275
currents 21, 153, 159, 163, 164, 341, 355
  deep **184–185**
  surface **182–183**
cyclones 273

D

dams 227, 229, 230, 233, 307
Dead Sea 120, 238
Death Valley 146
deciduous trees 322, 324
decomposers 317, 322, 324
deep water 184
deforestation 21, 319, 357
deltas 119, 217, 227, 229, 234
demersal zone 340
Denali, mountains 147
depressions 273, 274, 275
desertification 281, 319, 354, 355, 356
deserts 98, **122–123**, 136, 138, 140, 183,
  318, **332–333**
detritivores 317, 324
Devonian Period 23
dew 266, 270
dew point 261
diabase 37
diamonds 202, 296, 300, 302, *302*, 303,
  312
diatoms 188
dinosaurs 23, *26*
diorite 37
dip–slip faults 63
dolines 103
dolphins *340*
Drauchenhauchloch (cave) 120
drizzle 266

droughts 330, **354–355**
drumlins 127
Dust Bowl 281, 355
dust storms **280–281**

E

Earth,
  atmosphere 13, 32, **254–255**, 292
  chemical elements **32–33**
  core 33, 44, *45*, **46**
  crust 33, 38, 42, 44, *45*, 49, **50–51**, 52,
    60, 61, 69, 133
  formation **12–13**
  future Earth **364–365**
  geological periods **22–23**
  lithosphere **52–53**, 56, 81
  magnetic field 46, *47*, 60, 257, 299,
    365
  mantle 12, 33, 44, *45*, **48–49**, 52, 56,
    58
  Milankovitch cycles 349
  orbit 14, *16*, 17, 26, 288–289,
    288–289, 349
  shape **18–19**
  vanishing resources 296, **358–359**
earthquake zones 83, 85
earthquakes 21, 59, 60, 62, 67, **82–93**,
  94, 98, 191
  Benioff–Wadati zones 59
  damage **84–85**
  famous **92–93**
  measurement **86–87**
  prediction **90**
  seismic waves 33, 44, 45, 46, 52, *82*,
    83, **88–89**
East China Sea 233
East Pacific Rise 60, 61, 161
East Wind Drift 163
Easter Island 143, *160*
Eastern Highlands *see* Great Dividing
  Range

ecosystems 175, **316–317**, 320, 362
Eiger 205
El Niño 355
Elbert, Mount 200
Elbrus, Mount 134
electricity *207*, 230, 236, 239, 278, 307, 309
Ellesmere Island 151
eons 22
epicenter 83
epilimnion 184
Equator 19, 290, 322
Equatorial bulge 19
ergs 122, *123*
Erie, Lake 240, *240*
erosion 39, 42, 98, 99, 111, 114, 173, 197, 205, 281
eskers 127
estuaries 222, 224, 334, 338–339
ethanol 307, 310
Eurasia 150, 203
Eurasian tectonic plate 197, 204, 212
Europe 54, 132, **134–135**, 209
evaporation 164, 238, 260, 308, 332
Everest, Mount *66*, 67, 139, *196*, 197
exosphere 254, 255
extinctions 319, 321, 361, 363, 364
eye (of a hurricane) 286

F

factory farming 356
fangtooth fish *343*
farming 21, 227, 296, 308, 309, **310–311**, 347, 350, 355, **356–357**, 359
fast ice 156
fault blocks 67
fault zones 62
fault-slip 191
faults **62–63**
feldspar 35, 38

fens 119
fern frost 270
filter-feeders 341
Finke River 222
firestorms 85
firn 124, *125*
fish,
   coastal waters 339
   freshwater 334, 336
   oceans 341, 342, 343
Fish River Canyon 218
fishing industry 312, 313, 359
fissure volcanoes 74
fjords 338
flash floods 116, 123
floodplains 106, *107*
floods **116–117**, 119, 227, 347
fog **264–265**
fold mountains 59, 65
folds **64–65**
food chains and webs 316
forest floor 322, 324
forests 259
   boreal 318, **326–327**
   coniferous 138, 318, 325, 327
   rain forest 136, 140, 148, 230, 281, 316, 318, 320, 321, **322–323**, 325
   temperate **324–325**
   tropical 136, 318, 320, 321, **322–323**
fork lightning 278
fossicking 298
fossil fuels 35, **304–305**, 347, 352
   see also coal; gas; oil
fossil record 321
"fossil water" 246, 250
fossils 22, *28*, **28–29**, 42, 55, 145, 215
freezing point 270
freshwater 152, 197, 224, 242, 246, **308–309**, **336–337**, 359
frontal fog 265
frost 266, 270
frost heave 129

frost shattering 100
Fuji, Mount 67
Fujita scale 285

G

gadolinium 296
Galapagos Islands 320
Ganges River 197, **234–235**
garnet 302
gas 35, 36, 201, 203, 209, 296, 305, 312, 350
gemstones 203, 296, **302–303**
geodesy 18
geoid 18, 19
geology **22–23**, 36
geophysical surveys 299
geothermal energy 306, 358
geysers 81, *146*
giant tortoises *320*
Gibraltar *135*
Gilgit River 216
glacials 24, 25
glaciated landscapes **126–127**
glaciers 21, 25, 27, 39, 99, 111, 120, **124–127**, 128, 144, 148, 205, 211, 238, 241
glaciofluvial drift 127
Global Positioning System (GPS) 55, 56
global temperatures 26, 27, 346
global warming 127, 159, 305, **346–347**, 351, 365
gneiss 36, 41, 51
gold 33, 34, 199, 201, 203, 205, 296
Gondwana 132, *133*, 140
gorges *see* canyons and gorges
grabens 63
Gran Chaco 149
Grand Canyon 111, 147, **214–215**
granite 38, *39*, 51
grasslands 136, *136*, *138*, 140, 149, 318, **330–331**

gravel 297, 312
gravimetry 299
gravity 12, 17, 21, 44, 167, 299
Great Australian Bight 173
Great Barrier Reef 143, 175
Great Bear Lake **248–249**
Great Britain 151, 284
Great Dividing Range 141, **206–207**
Great Lakes 147, 238, **240–241**, 309
Great Pacific Garbage Patch 313
Great Rift Valley 63, 137, *137*, 246
Green Revolution 311
greenhouse effect 346, 347
greenhouse gases 27, 305, 310, 349
Greenland *150*, 151, 219
groundwater 90, 119, 129
growlers 176
groynes *172*
grykes 103
Gulf of Mexico 146
Gulf Stream 21, *182*, 183
gulfs 173
guyots 178
gyres 182, 183

H

habitat loss 319, 361
haematite 35, *300*
hail 266, **269**
halite 42
hamada 122
Hawaiian Islands 25, *80–81*, 81, *143*
hawksbill turtles *363*
Hayward Fault *91*
headlands 172
helicoidal flow 113
helium 254
herbivores 316
heterotrophs 316
high tides 167

hill spurs 127
hills **108–109**, 123
Himalayas 67, 69, 138, 139, **196–197**, 234
hoar frost 270
Holocene Period 27
Holocene Mass Extinction 361
Homo sapiens 136, 364
Honshu 151
Horn of Africa *50*
horst blocks 63
hot spots **80–81**, 147, 321
hot springs 81
Hudson Bay 173
humidity 261
Huron, Lake 240, *240*
hurricanes **286–287**
Hwang Ho 116
hydrocarbons 350
hydroelectricity *207*, 230, 236, 239, 307, 309
hydrothermal vents **180–181**, 342
hyperthermals 26
hypocenter 83
hypolimnion 185

I

ice 101, 124, 128, 196, 259, **270–271**, 308
ice ages **24–25**, 140, 365
ice caps 24, 25, 145, 346, 347
ice cores 27, 250, *251*, *347*, 349
ice crystals 262, 266, 268, *269*, 270, 278
ice islands 176
ice sheets 27, 128, *150*, 162, 211
icebergs 162, *163*, **176–177**
Iceland 212
ichthyosaurs *29*
icicles *271*
igneous rocks 22, 36, **38–39**, 40

index fossils 29
India 55, 139, 158
Indian Ocean 152, *153*, **158–159**, 175, 186
indicator species 334
Indo-Australian tectonic plate 140, 142, 143, 159, 197
Indus River Gorge **216–217**
Indus Submarine Fan 217
infrared (IR) light 258
Inga Falls 230
insolation 259
intensive farming 356, *357*
interglacials 24, 26, 27
internal deformation 125
International Ice Patrol 177
International Union for Conservation of Nature (IUCN) 360–361
involutions 128
iron 12, 32, 33, 35, 44, 46, 181, 199, 203, 205, 209, 296, 301, 302, 312, 359
irrigation 227, 235, 308, 311, 355
islands **150–151**, 155, 159, 161, 188, 212, 241
island arcs 59
    see also individual index entries
isobars *272*, 273
isostasy 69
Izu-Ogasawara Japan Trench 186

J

Jackson, Lake 239
jaguars *149*
Jakobshavn Glacier 127
Java Trench 158
Jeita Grotto 104
jet 302
jet streams 275
Jungersen Glacier 176
Jurassic Period 23, 26

K

Kaibab limestone 215
Kalahari 136
Kali Gandaki Gorge 218
Kani Bil River 222
Kariba, Lake 239
karst 103
Kazakhstania 203
kelp 339
Kermadec Trench 186
Kilimanjaro, Mount 67, 137
kimberlite 303
Kirkpatrick, Mount 211
Kiruna Mine 301
Kosciuszko, Mount 206
krill 341
Krubera Cave 104
krypton 254
Kuril-Kamchatka Trench 186

L

La Réunion 78, 266
Ladoga, Lake 134
lakes 119, **120–121**, 129
    formation **238–239**
    oxbow lakes *107*, 239, *239*
    see also individual index entries
land degradation 356, 357
landscape changes **98–99**
landslides 82, 94
lapilli 79
Late Heavy Bombardment (LHB) 13
Laurasia 132, *133*
lava 36, 38, 61, 74, 78, 81
lead 33, 35, 201, 209
lichen woodland 327
lightning 278, *279*
limestone 41, 42, 100, **102–103**, 104,
    215, 239, 297
Lion Cave 301
lithium 199

lithosphere **52–53**, 56, 81
Livingstone Falls 230
Lloro 266
longshore drift 173
love (Q) waves 89, *89*
lows 274, 276
Lualaba River 230, *231*
Lukunga River 246

M

macaws *323*
Mackenzie River 248
Madagascar 151, 321
Madeira River 222
magma *37*, 38, 41, *49*, 59, 61, 72, 73,
    74, 76, 77, 78, 180, 303
magma chambers 72, 73
magnesium 32, 33, 48
magnetic Poles 47, 145, 257
magnetism 46, 47, *47*, 60, 90, 257, 299,
    365
magnetometry 299
magnetosphere 47
major life zones *see* biomes
Maldives *158–159*, 159
Mammoth Cave 104
mammoths 29
manganese 312, *313*
mangrove swamps *118*, 119, 334
mantle plumes 44, 61, 80
marble *40*, 41, *41*
"mares' tail" clouds 262, *275*
Mariana Islands 188
Mariana Trench 58, 161, 178, 186,
    188–189
marine snow 178, 341, 342
marshes **118–119**, 334
martens *326*
meanders 106, *107*, 111, 113
medicines 312
Mediterranean climate 134, 290

Mediterranean Sea 164, *165*, 184, 227
Meghna River 234
Meiji Seamount 81
Melanesia 143
meltwater 111, 217, 233, 234, 238
Mercalli scale 86, 87
mercury 209
mesas *43*, 123
Mesoamerican Barrier Reef 175
mesosphere 254, 255
Mesozoic Era 22, 209
metals 35, 36, 296, **300–301**, 359
   see also individual index entries
metamorphic rock 36, **40–41**
meteorites 13, 33, 255
meteorology 292
meteosats 293
methane 27, 254, 310, 346, 347
Meuse River 222
mica 203
Michigan, Lake 240, *240*
Micronesia 143
Mid-Atlantic Ridge 60, 61, *154*, 155,
   212–213
mid-ocean ridges 60, 180, 195
Midway Island 161
migration, animal 327, 335
Milankovitch cycles 349
Milwaukee Depth 191
mineralogy 36
mineraloids 35
minerals 21, **34–35**, 36, 38, 199, 201,
   202, 203, 205, 209, 296, 297,
   300–301, 312
Mir diamond mine 303, *303*
Mississippi-Missouri river system 116, 147,
   201, **228–229**
Missouri River Valley 229
mist **264–265**, *269*
Moho Discontinuity 48
Mohs Scale 300
molybdenum 201

moment magnitude scale (MMS) 86
monsoon climate 290
monsoons 116, 159, 234
Mont Blanc 204, 205
montane forest 322
Moon 12, **16–17**, 167
moraine *125*, 127
mountain climates 290
mountains **66–67**, 108, 318, 325
   fold mountains 59, 65
mountain systems **194–195**
   peaks 67, 99
   ranges 67, **68–69**, 194, 195
   temperatures 67, 329
   underwater 20, 60, 61, 81, 161, 174,
      178, 186
   see also individual index entries
mudflats 113, 167
Murray-Darling river system 206

N

Nanga Parbat (mountain) 216
nappes 65
Narodnaya, Mount 203
Nasser, Lake 227
nature reserves and zoos 362
Nazca tectonic plate 199
neap tides 167
Neogene Period 23
neon 254
Nevado Misimi, Mount 225
névé 124, *125*
New Caldeonia Barrier Reef 175
New Guinea 140, 142, 143, 151, 195
New Orleans 228, *287*
New Zealand 25, 132, 142, 143, 195
Niagara Falls 115, *147*, *241*
nickel 12, 21, 33, 44, 46, 312
Nile perch 245
Nile River 116, 222, **226–227**, 309
nimbostratus clouds 266, 274, *275*

nimbus clouds 262
nitrogen 254, 257
nitrous oxide 254
North America 54, 132, **146–147**
North American tectonic plate 147, 191, 200, 212
North Atlantic Drift 134, 183
North European Plain 134
North Fork Roe River 223
North Pole *47*, 124, 268, 270, 328
Northern Hemisphere 152, 182, 184, 274, 276, 288, 289, 325
Northern Lights *256*, 257
Novaya Zimlya islands 134
nuclear power 307
nunataks 128, 211

O

Ob River 222–223, *223*
ocean trenches 58, 161, 178, *179*, **186–191**, 343
Oceania 132, **142–143**
oceanic climates 290
oceanic crust 51
oceans 19, 20, **152–153**, 347
  currents 21, 153, 159, 163, 164, **182–185**, 341, 355
  deep **178–179**, **342–343**
  life **340–343**
  resources, use of **312–313**
  temperatures 21, 27, 152
  see also individual index entries
oil 35, 36, 46, 65, 201, 203, 296, 305, 312, 347, 350, 358
Ojos del Salado, volcano 199
omnivores 316
Ontario, Lake 240, *240*
ooze 178, 188, 342
opals 35
Ordovician Period 23
ores 35, 36, 41, 300, 359

osmium 37
Otzi the Iceman 204, *204*
overfarming **356–357**
overfishing 312, 359
overgrazing 356
oxbow lakes *107*, 239, *239*
oxygen 32, 33, 35, 48, 50, 51, 67, 254, 257, 302, 305, 313
ozone 21, 346, 350, 351, 353, *353*

P

Pacific Mountain System 195
Pacific Ocean 60, 142, 143, 152, *153*, **160–161**, 175, 186, 199
Pacific tectonic plate 56, 142, 143, 188, 195
pack ice 156
Palaeozoic Era 22, 208
Paleocene-Eocene Thermal Maximum (PETM) 27
Paleogene Period 23
Pampas 149, 330
pancake ice 162
Pangaea 54, 132, *133*, 212
Pantanal 149
Panthalassa 54
Patagonia 149
pearls 302, 337
peat 119
pelagic zone 152, 340
Peléean eruptions 75
penguins *145*, *162*, 328
peridotite 36
periglacial conditions 128–129
Perito Moreno glacier *148*
permafrost 29, 128, 328
Permian Period 23, 321
Persian Gulf 159
Peru–Chile Trench 187
petrochemcials 296, 356
petrology 36

Petterman Glacier 176
phaneritic rocks 38
Philippine Trench 186
phosphate 359
photons 257
photosynthesis 339, 340, 365
photovoltaic cells 307
phytoplankton 338, 340
piedmont glaciers 125
pike *336*
Pikes Peak *201*
pillow lava 61
pingos 129
pitchblende 248
planets 12, 14
planetesimals 12
plants,
   coastal habitats 339
   deserts 332
   evolution 23
   oceanic 340
   vegetation succession 316
   wetlands 334, 337
   see also forests; grasslands
Platte River 201
Pleistocene Eemian interglacial 27
Pleistocene Ice Age 24
Pleistocene Period 27
Plinian eruptions 75
plunge pools 114, *114*
Point Nemo 143
polar climates 290
poljes 103
pollution 175, 235, 236, 265, 308,
   313, 319, **350–351**, 352, 359
Polynesia 143
pools 113
population growth 364
Port Radium 248, *249*
potassium 33
potholes 103, 104
prairies 318, 330, 331

Precambrian Eon 22, 23, 200, 208
precipitation 260, 266
   see also dew; frost; hail; rain;
     sleet; snow
prevailing winds 276, *277*
primary consumers 316
primary (P) waves 44, 88, *88*
primary producers 338
Pripet Marshes 119
proglacial lakes 239
Provo Canyon 219
Puerto Rico Trench 155, **190–191**
pumice 79
Puncak Jaya 143
pyroclasts 76, 79

Q

quartz 35, 38, 50, 205, 300
quasi-satellites 14
Quaternary Period 23
Quito 149

R

radiation fog 265
radio-dating 22, 23, 28
radon 90
rain 260, 262, **266–267**, 268, 269, 308
rain forests 136, 140, 148, 230, 281, 316,
   318, 320, 321, **322–323**, 325
Rakaposhi, mountain 216
rapids 106, 230
rare earths 359
Rayleigh (R) waves 89, *89*
red ochre 301
Red Sea *50*, 159
reg 122
regolith 109
relative humidity 261
renewable energy **306–307**, 312,
   358, 365

rhinos *360*
Richter scale 86
riffles 113
rift valleys 61, 63, 137, *137*, 242, 246, 248
rime 270
Ring of Fire 195, 199
Rio Grande 201
rivers **106–107**, 114, **222–223**
    channels 111, **112–113**
    deltas 119, 217, 227, 229, 234
    estuaries 222, 224, 334, 338–339
    underfit/misfit 111
    see also individual index entries
rock salt 42
rocks **36–37**
    crustal rock 50–51, 59
    dating 22, 23, 29
    erosion 39, 42, 98, 99, 111, 114, 173, 197, 205, 281
    faults **62–63**
    folds **64–65**
    geological column 22, 23
    igneous 36, **38–39**, 40
    mantle rock 48, 59
    metamorphic 36, **40–41**
rock cycle *36–37*
    sedimentary 36, 40, **42–43**, 204
    weathering 36, **100–103**, 104, 108, 109
rocky coasts **172–173**, 339
Rocky Mountains 146, **200–201**, 219
Rodinia 54
Rodrigues Triple Point 159
Romanche Trench *213*
Ross Ice Shelf 177
Rossby waves 275
rubidium 23
rubies 302
Russell Fjord 239
Ruwenzori Mountains 137

S

Sahara 23, 122, 136, 137, 281, 332
Sahel region 355
St Helens, Mount *76*, 81
salinity 153, 312, 341
salt 296
salt water 152, 164, 172, 184, 224
saltating 280
San Andreas Fault *62*, 63
sand dunes 122, *122*, 137, 215
sandbars 113
sandstones 42, 141
sandstorms **280–281**
sandy shores **170–171**, 339
sapphires 302
Sarawak Chamber 104
Sargasso Sea 155
satellite laser ranging (SLR) 55
satellites 18, 19, **20–21**, 44, 56, 90, 293, 299
savannas 136, *136*, 318, 330, *330*, 331
scandium 296
schist 41, 51
"Screaming Sixties" 163
sea fog 265
sea ice 156, *157*, 162, 184
sea levels 27
Sea of Marmara 164
seafloor spreading 49, 60–61, *61*, 159, 161, 212
seals 242, *243*
seamounts 20, 60, 61, 81, 161, 174, 178, 186
seas 152, **164–165**, 329
seasons 14, **288–289**
seaweeds 155, 339
secondary consumers 316
secondary (S) waves 44, 88, *88*
sedimentary rock 36, 40, **42–43**, 204
sediments 27, 36, 42, 59, 107, 215, 242, 250, 349

seiches 164
seifs 122, *122*
seismic gaps 90
seismic waves 33, 44, 45, 46, 52, *82*, 83, **88–89**
seismology 52, 90, 299
seismometers 86
Seram 140
Seychelles 159
shales 42
shallows 113
sheet lightning 278
shield volcanoes 74, *75*, 81
shingle 170, 173, 339
silicates 33, 35, 48, 50, 300
silicon 32, 33, 35, 48, 50, 51
silt 106, 120, 170, 229
Silurian Period 23
silver 199, 209, 296
silver tungsten 201
sinkholes 103, 239
slate 41
sleet 266
slip-off slopes 113
smog 265, 305, 351
snow 124, 196, 203, 259, 266, **268**, 269, 308
  blizzards **282–283**
snowdrifts 282
snowflakes 268
snowline 283
Snowy River 206
sodium 33
solar eclipses 17
solar power 307, 358
solar radiation 258–259, 365
Solar System 14
solar wind 257
Son Doong Cave 104
South America 54, 55, 132, **148–149**
South American plate 199, 212

South Pole *47*, 124, 144, 145, 270, 328
Southern Hemisphere 152, 182, 274, 276, 289
Southern Lights 257
Southern Ocean 152, *153*, **162–163**
Southwest Pacific Islands 132
space 255, 365
speleotherms 104
spits *172*, 173
spring tides 167
stacks 172, *172*, *173*
stalactites and stalagmites 104
steels 301
steppes 318, 330, 331
stone polygons 129, 330
storm beaches 170
storm surges 287
strata 42, 64, 65
stratocumulus clouds 262
stratosphere 254, *255*
stratovolcanoes *see* composite volcanoes
stratus clouds 262
streams 123
stromatolites 28
Strombolian eruptions 74
subduction zones 48, 49, 51, 60, 76, 178, 186, 191
subtropics 136
sugarcane 307, 310
sulfur 33, 36, 181
Sumatra 73, 151
Sun 14, 167, 273, 276, 290, 365
sunshine **258–259**
sunspots *348*, 349
supercells 284
Superior, Lake 240, *240*, 241
surveying and prospecting 36, **298–299**
swamps **118–119**, 215, 334
swell 168
synclines 65

T

taiga 326, 327
Tanganyika, Lake 120, **246–247**
Tasmania 25, 140
tectonic plates 52, **56–61**, 62, 64, 68, 81, 82, 83, 137, 139, 140, 159, 180, 186, 191, 238, 365
  subduction **58–59**, 178, 186, 188, 199
temperate climates 290
temperature extremes 138, 144, 146, 270
tension faults 62, 63
tephra 73, 79
Tethys Sea *133*, 197, 204
thermoclastis 101
thermoclines 185
thermohaline circulations 184
thermosphere 254, 255
Three Gorges 219, *219*, 233, 307
thunderclouds 262, 266, *267*, 275, 278, 284
thunderstorms **278–279**, 286
Tibetan Plateau *139*, 197
tidal bulges 167
tidal waves *see* tsunamis
tides 17, 161, 164, **166–167**, 172, 312, 338
till 127
tin 199, 301
titanium 33, 302
Titicaca, Lake 120
Tonga Trench 186
Tongariro National Park *142*
Tore-Madeira Rise *213*
Tornado Alley 285, *285*
tornadoes 275, **284–285**
Toubkal, Mount 208, *209*
trace fossils 28
trade winds 206
Transantarctic Mountains **210–211**
transform faults 57, 60

tree rings 27, 348
Triassic Period 23
tributaries 106, 217, 222, 224, 226, 230, 242
Tristan de Cunha 212
Tropic of Cancer 322
Tropic of Capricorn 322
tropical climates 290
tropics 136, 149, 266, 276, 290, 320
troposphere 254, *255*
tsunamis 86, **94–95**, 191
tundra 128, *128–129*, 318, 329
twisters *see* tornadoes
typhoons 286

U

ultraviolet (UV) light 254, 255, 258, 351, 353
Uluru *141*
unconformities 23, 215
understory 323, 324
upslope fog 265
Ural Mountains **202–203**
uranium 23, 33, 296
urban environments 319

V

valleys 63, 106, **110–111**, 126
  rift valleys 61, 63, 137, *137*, 242, 246
vegetation succession 316
Verkhoyansk, Siberia 138
Very Long Baseline Interferometry (VLBI) 55
Vesuvius, Mount 74, 75
Victoria Falls 115, *115*
Victoria Island 151
Victoria, Lake 137, 238, **244–245**
Vinson Massif 145
volcanic arcs 67
volcanic bombs *73*, 79

Volcanic Explosivity Index (VEI) 77
volcanoes 13, 59, 67, **72–81**, 149, 188, 199, 254
  active 77, 140, 199
  dormant 77, 140
  eruptions 72–73, *72–73*, **74–79**, 82, 94, 98, 349
  extinct 77, 137
  hot-spot volcanoes **80–81**, 147
  lava 36, 38, 61, 74, 78, 81
  magma *37*, 38, 41, *49*, 59, 61, 72, 73, 74, 76, 77, 78, 180, 303
  undersea 161, 174
Vostok 144, 270
Vostok, Lake **250–251**
Vulcanian eruptions 74

W

wadis 123
Wai-'ale-'ale, Mount 266
warm fronts 274, *275*
water cycle 308
water hyacinth 245, 337
water pressure 188, 342
water vapor 206, 254, 260, 261, 262, 264, 269, 270, 308
waterfalls **114–115**, 230, *271*
waterspouts *284*
wave-cut platforms 173
waves 164, **168–169**, 170, 172, 173
  tsunamis 86, **94–95**, 191
weather 290
weather forecasting **292–293**
weather fronts 265, 266, **274–275**
weathering 36, **100–103**, 104, 108, 109
West Antarctic Rift 211
wetlands **118–119**, 149, 318, **334–335**
whale falls 342–343
whaling 361
White Nile 226
whiteouts 282

wildebeest *330*
wildfires 330
Wilhelm, Mount 143
wind 163, 274, **276–277**
  hurricanes **286–287**
  tornadoes 275, **284–285**
wind power *306*, 307, 358
wind stress 168
"World Ocean" **152–153**
wrench faults 63

X

xenon 254

Y

Yangtze River 139, 219, *219*, **232–233**
Yarlung Tsangpo Gorge 218
Yellow River 116
Yellowstone National Park 81, *146*, 147
Yenisei River **236–237**, 242

Z

zebra *330*
zinc 21, 33, 35, 181, 201
zooplankton 340
zoos *see* nature reserves

Acknowledgments

All artwork from the Miles Kelly Artwork Bank

The publishers would like to thank the following sources for the use of their photographs:
t = top, b = bottom, l = left, r = right, c = centre

Front cover Momatiuk-Eastcott/Corbis
Back cover (t) Catmando/Shutterstock.com, (c) Atiketta Sangasaeng/Shutterstock.com,
(b) Galyna Andrushko/Shutterstock.com

Alamy 207 Bill Bachman; 226–227 Peter Barritt; 231 Images of Africa Photobank;
244 Peter Horree; 247 Universal Images Group Limited; 358 Westend61 GmbH

Corbis 15 Bettmann; 121 NASA/Earth Observatory/Handout; 187 Xinhua/Xinhua Press;
208–209 Yann Arthus-Bertrand; 360 Cyril Ruoso/ JH Editorial/Minden Pictures

Dreamstime 115 Fabio Cardano; 118 Lawrence Wee; 176–177 Vladimir Seliverstov; 256

Fotolia.com 28 Ismael Montero; 66 QiangBa DanZhen; 144–145 steve estvanik;
166 overthehill; 173(r) Michael Siller; 284 KoMa; 352 sisu

Frank Lane Picture Agency (FLPA) 322–323 Murray Cooper/Minden Pictures; 363 Reinhard
Dirscherl; 62 Kevin Schafer/Minden Pictures; 65 Michael & Patricia Fogden/Minden Pictures

Getty Pictures 84 Martin Bernetti/AFP

Glow Images 141 Vidler Steve; 150 Bert Hoferichter; 198 Wayne Lynch;
218–219 View Stock; 249 Jason Pineau

iStock.com 34–35 robas; 78–79 titine974; 102 graphicjackson; 136 andydidyk; 318 naes;
354 cinoby

NASA 18 Apollo 17 Crew; 20 USGS EROS Data Center; 50 SeaWiFS Project, GSFC/
ORBIMAGE; 59; 91 ESA; 106 Jesse Allen, courtesy of the University of Maryland's Global
Land Cover Facility; 137 NASA/JPL/NIMA; 139 Jeff Schmaltz, MODIS Rapid Response Team,
GSFC; 182 Norman Kuring, MODIS Ocean Team; 240; 286; 287; 348 GSFC Scientific
Visualization Studio

NOAA 154; 180–181

Rex Features 188–189 KeystoneUSA-ZUMA; 202 Keith Waldegrave/Associated Newspapers;
210 NASA/Michael Studinger; 237 Russian Look/UIG; 364 Solent News; 347 Nick Cobbing

Science Photo Library 16 Gary Hincks; 26 Roger Harris; 49 Claus Lunau; 190 NOAA; 194 Gary Hincks; 213 Gary Hincks; 238 Gary Hincks; 251(t) Nicolle Rager-Fuller, National Science Foundation; 272–273 Karsten Schneider; 298 Simon Fraser; 303 Ria Novosti; 313 Institute of Oceanographic Sciences/NERC; 353 NASA/Goddard Space Flight Center; 355 NASA

Shutterstock.com 1 Vladimir Melnikov; 10–11 yexelA; 24–25 Incredible Arctic; 30–31 George Burba; 39 2009fotofriends; 40 Ververidis Vasilis; 41 Olena Tur; 43 Sebastien Burel; 47 Snowbelle; 68–69 Andrey_Popov; 70–71 beboy; 92 arindambanerjee; 96–97; 98–99 Xavier Marchant; 100–101 Patrick Poendl; 105 pmphoto; 108–109 Robert Plotz; 110 RIRF Stock; 112 Ulrich Mueller; 117 Asianet-Pakistan; 123 ricardomiguel.pt; 126 vrihu; 128–129 George Burba; 130–131 tororo reaction; 133 Designua; 138 Daniel Prudek; 142 Pichugin Dmitry; 146 Lee Prince; 148 robert cicchetti; 149 Andre Dib; 153 ktsdesign; 155 Roman Sulla; 157 CatchaSnap; 158–159 Paolo Gianti; 160 vasen; 163 kkaplin; 165 Joao Virissimo; 173(l) Robyn Mackenzie; 174 Brian Kinney; 175 aquapix; 192–193 prochasson Frederic; 196–197 Dchauy; 200–201 John R. McNair; 205 Fedor Selivanov; 214–215 Francesco R. Iacomino; 216–217 siraphat; 220–221 Pichugin Dmitry; 223 Ice Cherry; 224–225 Johnny Lye; 228 Ed Metz; 232–233 kanate; 234–235 Radiokafka; 239 Javier Rosano; 241 Gary Blakeley; 243 withGod; 252–253 Corepics VOF; 258–259 Paul Aniszewski; 260–261 L.M.V; 263 Andrzej Gibasiewicz; 267 dpaint; 271 evronphoto; 279 André Klaassen; 280 Andrew McConnell; 282–283 Vadym Zaitsev; 292–293 George Burba; 302 RTimages; 294–295 Jaochainoi; 296–297 curraheeshutter; 300–301 bondgrunge; 304 russal; 306 Rodrigo Riestra; 308 Chailalla; 311 Kletr; 314–315 Alan Uster; 319(l) EcoPrint, (r) Mikadun; 320–321 Ryan M. Bolton; 324–325 Jacob Whyman; 326 Steve Brigman; 328–329 Andrea Izzotti; 330–331 Oleg Znamenskiy; 332–333 Anton Foltin; 335 Brian Lasenby; 336–337 Kletr; 338–339 twospeeds; 340–341 Krzysztof Odziomek; 344–345 Galyna Andrushko; 350–351 Ian Bracegirdle; 357 Rihardzz